Compendium of
Bioenergy Plants
SUGARCANE

Compendium of Bioenergy Plants

Series Editor

Chittaranjan Kole
Dean, Jacob School of Biotechnology & Bioengineering
Sam Higginbottom Institute of Agriculture,
Technology & Sciences
Formerly Allahabad Agricultural Institute
Allahabad, UP
India

Books in this Series:

- Stephen L. Goldman & Chittaranjan Kole: *Corn*
- Hong Luo, Yanqi Wu & Chittaranjan Kole: *Switchgrass*
- Eric Lam, Helaine Carrer, Jorge A. da Silva & Chittaranjan Kole: *Sugarcane*

Compendium of Bioenergy Plants
SUGARCANE

Editors

Eric Lam
Distinguished Professor
Plant Biology & Pathology
Rutgers, The State University of New Jersey
New Brunswick, New Jersey, USA

Helaine Carrer
Department of Biological Sciences
Agriculture College "Luiz de Queiroz"
University of São Paulo
Piracicaba, Brazil

Jorge A. da Silva
Texas A&M AgriLife Research – Weslaco
Soil & Crop Science Department
Texas A&M University System
Weslaco, Texas, USA

Chittaranjan Kole
Dean, Jacob School of Biotechnology & Bioengineering
Sam Higginbottom Institute of Agriculture,
Technology & Sciences
Allahabad, India

CRC Press
Taylor & Francis Group
Boca Raton London New York

CRC Press is an imprint of the
Taylor & Francis Group, an **informa** business
A SCIENCE PUBLISHERS BOOK

CRC Press
Taylor & Francis Group
6000 Broken Sound Parkway NW, Suite 300
Boca Raton, FL 33487-2742

First issued in paperback 2019

© 2016 by Taylor & Francis Group, LLC
CRC Press is an imprint of Taylor & Francis Group, an Informa business

No claim to original U.S. Government works

ISBN-13: 978-1-4987-4338-9 (hbk)
ISBN-13: 978-0-367-37728-1 (pbk)

**Visit the Taylor & Francis Web site at
http://www.taylorandfrancis.com**

**and the CRC Press Web site at
http://www.crcpress.com**

Preface to the Series

The need for sustainable energy is growing at an increasing rate with the alarmingly high rate of increase in population coupled with the fast growth of urbanization. By 2050 the world population is estimated to be seven billion computed at a conservative rate of growth. By 2100, the number is projected to be over ten billion by another estimate. The source of fossil fuels being predominantly used over time will face depletion around the end of this century unless non-conventional energy sources are put in place. Besides depletion, fossil fuel use is constrained by geo-political issues and threat of greenhouse gas emission. Among the alternative energy sources, bioenergy is emerging as the most promising as compared to atomic, solar, and wind. Bioenergy including bioethanol and biodiesel can be produced from cellular biomass, starch, sugar, and oil derived from several plants and plant products in huge amounts once the required strategies and technologies are formulated and validated for commercialization in cost-effective ways.

Scientific exploratory research conducted during the last few years has identified a large number of plants as potential sources of bioenergy. These include maize, sorghum, switchgrass, canola, soybean, and sugarcane among field crops; eucalyptus and poplar among forest trees; and jatropha, oil palm and cassava among plantation crops. Several other promising field crops including Brachypodium, minor oilseeds, sugarbeet, sunflower, and sweetpotato; forest trees including diesel trees and shrub willow; plantation crops such as Paulownia; many lower plants; and even vegetable oils, organic farm waste and municipal sludge have been found to be promising. Therefore, 'fuel' has made its place in the list of principal agricultural commodities along with food and fiber.

Significant studies have also been conducted in natural and social sciences to facilitate utilization of plants and plant products as the most potential source of bioenergy. In bioenergy crops, research has been carried out on genetics, genomics and breeding for relevant traits employing traditional and molecular breeding, genomics-assisted breeding, and genetic engineering. Physiological works have been done for *in planta* production of cell-degrading enzymes and enzymatic conversion of cell walls into biofuels. Significant advancement has been made on the works on post-harvest technologies and chemical engineering, fuel quality, and greenhouse gas impacts of bioenergy. Most importantly, economics, public policies, and perceptions have also been critically examined.

There are, at present, only a few books on bioenergy crop plants available. I have myself edited a book recently with two other co-editors for the CRC Press of the Taylor and Francis Group. This book entitled 'Handbook of Bioenergy Crop Plants' elucidates on the general concepts of and concerns about bioenergy crop production, genetics, genomics and breeding of commercialized bioenergy crop plants, and emerging bioenergy crops or their groups besides deliberations on unconventional biomass resources such as vegetable oils, organic waste and municipal sludge.

As expected, there is also an array of research and review articles on the basic concepts, strategies and means of utilization of bioenergy crop plants and their products in scientific journals, web sites, newsletters, newspapers, etc. However, there is no endeavor to present any compilation about all the relevant aspects related to particular bioenergy crop plants already commercialized or having potential to be commercialized in near future. The present book series will hopefully fill up that vacuum. This is particularly important as the subject of bioenergy has already occupied its place in academia, research labs, and public life. This was the underlying force behind conception of a book series on 'Compendium of Bioenergy Crop Plants'.

At the outset, I formulated the tentative outline for 15 chapters to maintain more or less uniformity throughout the volumes of the compendium. These included basic information on the crops; anatomical and physiological researches relevant to feedstock; special requirements related to agricultural and industrial infrastructure; elucidation on genetics, genomics and breeding of bioenergy traits; public platforms for sharing results and building initiatives; role of public and private agencies in fostering research and commercialization; regulatory, legal, social and economic issues; general concerns and their compliance; and also future prospects and recommendations. However, the volumes of this compendium are devoted to various crop plants and obviously the concerned volume editors had to improvize on the contents of the respective volumes based on the unique information available and specific requirements. Thus, each volume of this compendium has the 'stand-alone' potential at the same time, thanks to the excellent balancing job performed by the volume editors. Fortunately most, if not all, of the volume editors have long standing association with me as an author of a chapter in some other book, or volume editor of another book series or colleague in a research platform. Therefore, it has been highly comfortable and enriching for me to work with them again for this compendium. I take this opportunity to express my heartiest gratitude to them for offering me this opportunity. The authors of the chapters for each of the volumes have produced high quality deliberations both in terms of comprehensive contents and lucid write-ups. As the series editor, I must join with my volume editors to extend our thanks to the authors of the chapters for their elegant contributions as well as sincere cooperation all along.

This compendium was originally conceptualized by my wife and colleague, Phullara. She had meticulously reviewed the relative importance and quantum of works accomplished in the commercialized and promising bioenergy crops plants and had eventually identified the leading bioenergy crop plants to which the individual volumes of this compendium are devoted to. She was always there for help in editing this compendium similar to several other book series containing over sixty books published or in press. Expressing just thanks will not do justice to her contribution to this book project. I have, therefore, dedicated this compendium to her in recognition of her contributions to this book project and also for all her support, advice and inspiration for all my academic activities besides shouldering most of our domestic loads, taking the major responsibility to navigate our family and nourish our three growing kids, Papai, Titai and Kinai, as that provided me with enough extra time for my book editing jobs in addition to my professional duties.

Chittaranjan Kole

Dedication by Series Editor

*Dedicated to
My beloved wife and colleague,
Phullara*

*The infinite source of support, strength, guidance, and inspiration for my
mission to serve science and society.*

Preface

The 21st Century will be a pivotal one for human evolution as a species. While the previous 100 years witnessed two World Wars and the escalating power of human's destructive capabilities that culminated in nuclear weapon proliferation, the present century has brought us face-to-face with the consequences of human ingenuity. While we have averted rash actions during the Cold War that could have led to mass destruction by nuclear weapons, the rapid advances in our technologies that tap the resources of this planet ever more efficiently and rapidly is changing our global systems at an unprecedented scale. Although massive changes to earth's systems have been caused by life forms before, such as the production of oxygen in the atmosphere after the appearance of photo-oxygenic organisms, it has never occurred at a rate similar to what has been happening during the last two centuries. Two interrelated factors that helped to drive the massive changes in climate that are now widely recognized by the scientific community are sustained increase in global population as well as their economic wealth, and the transition to fossil fuels as the primary energy sources to power human societies across the globe. A quantifiable result of these factors is a steady rise in greenhouse gases, mainly carbon dioxide, in our atmosphere, which directly contribute to global warming. The negative impacts of these factors on our planet are most obvious in drastic swings in weather patterns with steady increase in ocean temperature, rapid rise of sea levels from melting of glaciers worldwide, and accelerated loss of species estimated to be 100 fold faster than 2 centuries ago.

With the recognition that fossil fuels are not inexhaustible and that their massive uses has severe consequences on the global climate systems, the beginning of the 21st Century saw a surge of interest on research and deployment of various renewable energy options. Success of at least some of these potential candidates to replace a significant portion of the global energy portfolio would raise hope that we may be able to mitigate the threat of global Climate Change by adopting new sources of renewable energy and fuels.

Currently, the most abundant energy that can be readily tapped on earth is light from the sun. Compared to global energy consumption rate, the rate of energy equivalent from solar irradiation reaching the earth is about 10,000 times. Thus, less than one hour of sunlight will be able to power all of our energy needs in a year if the energy from that irradiation can be captured efficiently. While advances in photovoltaic devices have steadily been improving, we believe that diversifying some of our existing agriculture systems toward renewable energy and fuel production can replace a significant percentage of our fossil fuel needs. This Compendium of Energy Crops endeavor to contribute to this effort by providing a comprehensive collection of papers from experts that are working in the frontline to facilitate the rapid optimization of various crop plants as feedstocks for fuel and energy production. This volume focuses on sugarcane, arguably the current energy crop that has the most significant impact on mitigating greenhouse gas footprint from energy consumption. The energy output to input ratio for sugarcane bioethanol has been estimated to be 8 to 10, while that for corn is currently around 1.4 as a comparison. With steady research and development over the past 50 years, sugarcane as an energy crop has been steadily improved through breeding approaches that resulted in varieties with higher sugar content, more resistant to diseases, and more optimal physical structure for easier harvesting and processing. The parallel development of Flex-fuel engine technology also improved consumer acceptance of bioethanol as an alternative to gasoline, especially in Brazil where government incentives and policies have played key roles in the success of their biofuel economy.

The opening chapter by Carneiro and colleagues provides a detailed description of the origin, genetic diversity and historical aspects of sugarcane and its domestication history from past to present. It provides a comprehensive account of all the basic information on sugarcane as a crop plant and its relation to global agriculture. Park and da Silva follow with the second chapter that describes in depth the current challenges in genetic improvement of sugarcane through hybridization with other related grasses. The challenges and perspectives described by the authors in this breeding approach would be very useful for readers who are interested in the development of new varieties of sugarcane with improved traits in stress tolerance and growth properties. While sugar that can readily be extracted from sugarcane has been used to produce bioethanol fuel, the remaining biomass composed of leaves and tops, usually left in the field, and the left-over stalks after crushing at the mill, called bagasse, is rich in cellulosic material that can also be converted to fuel with additional processing. To optimize this so-called second generation biofuel production from leaves and bagasse, knowledge of the sugarcane cell wall's chemistry and structure would be essential. Buckeridge and his coauthors provided an authoritative description of the complex composition and properties of the cell wall of sugarcane. The narrative especially emphasizes on the important factors that would need to be considered for optimal strategies to hydrolyze the cell wall components of sugarcane. It will serve as an important guide to readers who want to increase the amount of biofuel that can be produced from harvested biomass by more complete usage of all the energy-storage components that it contains. In addition to technology improvement for feedstock production and processing, it is also important to understand the economic forces in the marketplace that can determine the viability of the different products that can be produced from a crop such as sugarcane. To this end, Ribera and Bryant provide a concise analysis of the various government instruments in the United States that have created an environment that is in favor of sugar production rather than bioethanol. Without drastic changes in the tariff and other financial programs that protect the sugar industry in the U.S. market, sugarcane in the U.S. will be unlikely to be used for bioethanol. This chapter thus explains why in spite of the existence of large sugarcane production capacity in the U.S., bioethanol is chiefly, if not entirely, produced from corn starch. Looking toward the future demand that would need to be satisfied if renewable biofuels are to replace a major portion of the current fossil fuel production, traditional breeding strategies would be too slow to generate new elite varieties to improve on the quality and production reliability of sugarcane. In order to complement this approach, Barboza and Carrer discuss the technology of genetic transformation of sugarcane in terms of its current state-of-the-art as well as future areas for improvements. The authors provided a detailed description of the methodology, the successful traits that have been introduced into sugarcane so far in the past decade, and the new approaches that may overcome previous pitfalls and limitations. It should serve as a useful reference for new research or commercial groups who desire to use this technology to characterize genes and their associated traits in sugarcane and to develop varieties with novel phenotypes. To finish with our story on sugarcane and its relationship with human societies, it is important to bear in mind the historical setting in which it has been domesticated. In Brazil and elsewhere in the world, sugarcane plantations traditionally employed very low income sectors of the population and in earlier times, through slaves that worked under their owners. As the farming of sugarcane becomes more mechanized and health concerns dictated dramatic changes in the harvesting practice, major shifts in the labor-owner relationship has to be managed appropriately in order to make benefits to both. Moraes and Oliveira made an in-depth discussion of the past practice and current reforms that are taking place in the labor and societies that have close relationship to sugarcane and bioethanol production in Brazil. The unfolding story serves as a reminder of the complexity in implementing change in a well-established industry and its potential harm as well as benefits to the social dimensions, depending on how well this change is being managed.

In assembling this volume together, we hope that it can reach a wide audience to provide them with a comprehensive appreciation of the potential for this remarkable crop plant, its current impact on the fledgling renewable energy sector, and the promise that it holds to help build a more sustainable biofuel industry.

Eric Lam
Helaine Carrer
Jorge A. da Silva
Chittaranjan Kole

Contents

List of Contributors

André Luiz Barboza
Department of Biological Sciences, Escola Superior de Agricultura "Luiz de Queiroz", Universidade de São Paulo, Av. Pádua Dias, 11. Piracicaba-SP. 13418-900. Brazil.
Email: andrehlb@usp.br

Henry Bryant
Texas AgriLife Extension Service, Texas A & M University, 2401 East Highway 83, Weslaco, TX 78596.
Email: h-bryant@tamu.edu

Marcos S. Buckeridge
Laboratory of Plant Physiological Ecology (LAFIECO), Department of Botany, Institute of Biosciences, University of São Paulo.
Email: msbuck@usp.br

Monalisa Sampaio Carneiro
Universidade Federal de São Carlos, Centro de Ciências Agrárias, Rodovia Anhanguera, km 174 - SP-330, Araras 13600-970, São Paulo, Brazil.
Email: monalisa@cca.ufscar.br

Helaine Carrer
Department of Biological Sciences, Escola Superior de Agricultura "Luiz de Queiroz", Universidade de São Paulo, Av. Pádua Dias, 11. Piracicaba-SP. 13418-900. Brazil.
Email: hecarrer@usp.br

Jorge A. da Silva
Texas A&M AgriLife Research, The Texas A&M University System, 2415 East Highway 83, Weslaco, TX 78596, USA.
Email: jadasilva@tamu.edu

Márcia Azanha Ferraz Dias de Moraes
Departamento de Economia Administração e Sociologia, Universidade de São Paulo, ESALQ, Av. Pádua Dias, 11, 13418900 - Piracicaba, SP, Brasil.
Email: mafdmora@gmail.com

Fabíola Cristina Ribeiro de Oliveira
Departamento de Economia Administração e Sociologia, Universidade de São Paulo, ESALQ, Av. Pádua Dias, 11, 13418900 - Piracicaba, SP, Brasil.
Email: fbcoliveira@hotmail.com

Amanda P. de Souza
Laboratory of Plant Physiological Ecology (LAFIECO), Department of Botany, Institute of Biosciences, University of São Paulo.
Email: amanda.psouza@gmail.com

Hermann Paulo Hoffmann
Universidade Federal de São Carlos, Centro de Ciências Agrárias, Rodovia Anhanguera, km 174 - SP-330, Araras 13600-970, São Paulo, Brazil.
Email: hermann@cca.ufscar.br

Guilherme Rossi Machado Junior
G. Rossi Consulting Co., Piracicaba, 13400-123 São Paulo, Brazil.
Email: rossi@canevarieties.com

Jong-Won Park
Texas A&M AgriLife Research, The Texas A&M University System, 2415 East Highway 83, Weslaco, TX 78596, USA.
Email: jwpark@ag.tamu.edu

Luis A. Ribera
Texas AgriLife Extension Service, Texas A & M University, 2401 East Highway 83, Weslaco, TX 78596.
Email: lribera@tamu.edu

Wanderley D. dos Santos
Laboratory of Plant Biochemistry, University of Maringá, PR, Brazil.
Email: wanderley.dantasdossantos@gmail.com

Marco A.S. Tiné
Núcleo de Fisiologia e Bioquímica de Plantas, Instituto de Botânica de São Paulo.
Email: marco.tine@gmail.com

Sugarcane—Basic Information on the Plant

Monalisa Sampaio Carneiro,[1,]* *Guilherme Rossi Machado Junior*[2] *and Hermann Paulo Hoffmann*[1]

ABSTRACT

Sugarcane is a monocotyledonous, semi-perennial crop, allogamous and belongs to the Family Poaceae (grass family), and Genus *Saccharum*. The taxonomy and nomenclature of the genus has always been challenging. The original classification proposed for the genus by Linnaeus and nomenclature of the genus *Saccharum* has been revised by some authors. Traditionally, cultivated sugarcane is an important crop to produce sugar for human consumption. In recent years, economic interest in sugarcane has increased significantly due to the increased demand worldwide for sustainable energy production, like biofuels. Additionally, in the context of green chemistry, sugarcane has one of the best profiles as a carbon source. Furthermore, it is a source of renewable resources and

[1] Universidade Federal de São Carlos, Centro de Ciências Agrárias, Rodovia Anhanguera, km 174 - SP-330, Araras 13600-970, São Paulo, Brazil.
  Email: hermann@cca.ufscar.br
[2] G. Rossi Consulting Co., Piracicaba, 13400-123 São Paulo, Brazil.
  Email: rossi@canevarieties.com
* Corresponding author: monalisa@cca.ufscar.br

can easily be processed in nature or later under industrial conditions for generation of multiple products. In this chapter, we will summarize the current knowledge on the history, origin, genetic complexities and economics related to sugarcane as a domesticated crop plant. While sugarcane is an efficient model as a biofuel feedstock, some important challenges must be overcome for its further optimization as well as improvement. Chief among these is the genetic base of sugarcane germplasms needs to be broadened with the aim of increasing the genetic variability for genotypes with higher biomass and cold/drought tolerance in order to improve reliability of production of this crop in the face of global Climate Change. To this end, it will be necessary to improve the efficiency of molecular markers and transgenic technologies in order to expand the use of this knowledge to complement traditional breeding approaches.

Botanical Descriptions

Sugarcane is a monocotyledonous crop and allogamous, and belongs to the Order Poales, Family Poaceae (grass family), Subfamily Panicoideae, Tribe Andropogoneae, Subtribe Saccharinae and Genus *Saccharum*. This tribe includes tropical and subtropical grasses like sorghum and maize. The taxonomy and phylogeny of sugarcane is complicated with five genera that form closely related interbreeding groups known as the '*Saccharum* complex'. The '*Saccharum* complex' comprises *Saccharum*, *Erianthus* section *Ripidium*, *Miscanthus* section *Diandra*, *Narenga* and *Sclerostachya*. These genera are characterized by high level of polyploidy and frequently unbalanced numbers of chromosomes, making it difficult to determine the taxonomic relationships (Daniels and Roach 1987; Sreenivasan et al. 1987; Amalraj and Balasundaram 2006).

The original classification of Linnaeus and nomenclature of the genus *Saccharum* has been revised (Buzacott 1965; Daniels and Roach 1987; D'Hont and Layssac 1998). The *Saccharum* genus traditionally (modern sugarcane cultivars) comprises six species: *S. officinarum*, *S. sinense*, *S. spontaneum*, *S. robustum*, *S. edule* and *S. barberi* (D'Hont et al. 1998). However, Irvine (1999) has proposed that the genus should be reduced to two species: a) the species *S. officinarum* grouping together *S. sinense*, *S. robustum*, *S. edule*, *S. barberi* and *S. officinarum*; and b) *S. spontaneum* as traditionally defined. His proposal was based on: a) the unique basic chromosome number and distinctive DNA fingerprints of *S. spontaneum* from the other species of *Saccharum*, and b) interfertility of the grouped species and the lack of distinct characteristics to separate them into individual species.

Currently, the genus *Saccharum* is composed of the following species according to Brandes (1958).

Saccharum spontaneum L.

Chromosome number in *S. spontaneum* varies from 40 to 128 with a basic chromosome number of eight, and five major cytotypes: 2n = 64, 80, 96, 112, and 128 (Panje and Babu 1960; Sreenivasan et al. 1987; D'Hont et al. 1996). India is the centre of origin and diversity, with wide geographic distribution from New Guinea to Africa and the Mediterranean (Mukherjee 1957; Roach and Daniels 1987). Known as wild sugarcane, *S. spontaneum* L. is highly polymorphic, with plant stature ranging from small grasslike plants without stalks, to plants over 5 m high with long stalks. In breeding programs, *S. spontaneum* is the species that has contributed to increased disease resistance, tillering, yields, and ability to regrow after stems are cut (ratooning ability), but it has low sucrose content (Mohan Naidu and Sreenivasan 1987).

Saccharum officinarum L.

The vast majority of *S. officinarum* clones display 2n = 80 chromosomes, with a basic chromosome number of ten, making this species octoploid (Bhat et al. 1962; Sreenivasan et al. 1987). *S. officinarum* is a complex hybrid of different species as it is an autopolyploid and also an allopolyploid, with whole chromosomes in *S. officinarum* that are homologous with those in the genera *Miscanthus* and *Erianthus* section *Ripidium* (Sreenivasan et al. 1987; Daniels and Roach 1987; Besse et al. 1997). *S. officinarum* L. is known as the noble cane because it produces abundant sweet juice. It has thick stalks with low fiber content.

Saccharum robustum Brandes & Jesw. ex Grassl

S. robustum largely encompasses clones with 2n = 60 or 80 (Bhat et al. 1962), but also includes many other forms that may have up to 200 chromosomes (Price 1965). It is believed that it is the ancestral species from which *S. officinarum* is derived (D'Hont et al. 1998; Brown et al. 2007). It possesses a high fiber content and vigorous stalks that are 2.0–4.4 cm in diameter and up to 10 m tall. The culms are hard and have little juice, are poor in sugar content and have a hard rind, a characteristic that is exploited to build hedges (Bakker 1999).

Saccharum edule Hassk.

Edule clones have edible aborted inflorescence and have low sugar content (Grivet et al. 2004; Amalraj and Balasundaram 2006). Barberi clones have 2n = 60 to 80 chromosomes (Roach 1972) and are morphologically similar to *S. robustum* except that the inflorescence is compacted. The origin of the *S. edule* is not yet defined. Brandes et al. (1938) proposed it to be a mutant form of *S. robustum*. Grassl (1967) thought that *S. edule* of New Guinea evolved from *S. robustum* and *Miscanthus* and the Fiji form from *S. officinarum* and *Miscanthus*.

Saccharum sinense Roxb.

Chromosome number in *S. sinense* varies from 81 to 124 chromosomes. Sinense clones are tall plants, presenting long, green-bronze canes, fibrous stalks and broad leaves, and they are cultivated in China. *S. sinense* are thought to be derived from *S. officinarum* x *Miscanthus* introgression. Hemaprabha and Sree Rangasamy (2001) supported the *S. spontaneum* based origin of *S. sinense*. However, recent studies showed that *S. sinense* is a natural hybrid of *S. officinarum* and *S. spontaneum* (Irvine 1999; D'Hont et al. 2002). Some taxonomists consider *S. sinense* and *S. barberi* a single species.

Saccharum barberi Jesw.

Barberi clones have 2n = 111 to 120 chromosomes and are cultivated in northern India, with better adaptation to sub-tropical environments. Barberi clones are short with thin, cylindrical, grey green/white or ivory canes and narrow leaves and lower sugar content than Noble canes (Amalraj and Balasundaram 2006). The plants have a vigorous, well-developed root system and good tillering, which enables adaptation to poor and dry soil, as well as allowing the production of a large amount of biomass (Cheavegatti-Gianotto et al. 2011). According to Parthasarathy (1946), *S. barberi* evolved from *S. spontaneum* and *S. officinarum*. However, Grassl (1977) suggested that *Erianthus procerus* and *Sclerostachya* contributed to Barberi clones. Genomic and molecular cytogenetic data provided strong evidence that *S. barberi* was derived from interspecific hybridization between *S. officinarum* and *S. spontaneum* (D'Hont et al. 2002).

Center of Origin, Botanical Origin and Evolution, Domestication, Dissemination and Morphology

The genus *Saccharum* probably originated before the continents assumed their current shapes and locations. The genus consists of 35–40 species and has two centers of diversity: the Old World (Asia and Africa) and the New World (North, Central and South America) (Cheavegatti-Gianotto et al. 2011). Commercial sugarcane cultivars have arisen through intensive selective breeding of species within the *Saccharum* genus, primarily involving crosses between *S. officinarum* and *S. spontaneum* (Roach and Daniels 1987; Cox et al. 2000; Lakshmanan et al. 2005).

S. officinarum has been cultivated since prehistoric times (Barnes 1974; Sreenivasan et al. 1987). The hypothetical origin of *S. officinarum* from domestication of *S. robustum* in New Guinea to display high sugar and low fiber is accepted by most sugarcane breeders (Daniels and Roach 1987). The centre of origin for *S. officinarum* is thought to be in New Guinea, the Malayan Archipelago, or the Melanesian and Polynesian islands (Mukerjee 1957). The species was probably transported throughout Southeast Asia by humans, where a modern center of diversity was created in Papua New Guinea and Java (Indonesia), and specimens were collected in the late 1800s (Daniels and Roach 1987). From its origin, the sugarcane plant has experienced a wide dispersion as it accompanied humans in their various migrations. The varieties were brought westward in fourth century BC and subsequently grown in Persia, Arabia, and Egypt, thus forming the basis of the Mediterranean sugar industry. In the fifteenth century, a new Indian variety (Puri, later called "Creole" cane) was introduced and transferred to Madeira, the Canaries, Cape Verde Islands, São Tomé, and parts of West Africa. Beginning from 1493, "Creole" cane was planted in Brazil and the Caribbean until the mid-eighteenth century when varieties imported from the Old World began to replace it (Galloway 1989).

The center of origin and diversity of *S. spontaneum* is believed to have evolved in southern Asia. *S. spontaneum* can be grown in a wide range of habitats and altitudes (in both tropical and temperate regions) and it is spread over latitudes ranging from 8°S to 40°N in three geographic zones: a) east, in the South Pacific Islands, Philippines, Thailand, Malaysia, Taiwan, Japan, China, Vietnam and Myanmar; b) central, in India, Nepal, Bangladesh, Sri Lanka, Iran, Pakistan, Afghanistan, and the Middle East; and c) west, in Egypt, Kenya, Uganda, Tanzania, Sudan, and other Mediterranean countries (Daniels and Roach 1987; Tai and Miller 2001). The origins of both *S. sinense* and *S. barberi* seem to be accepted as being from hybrids of locally cultivated *S. officinarum* and locally growing *S. spontaneum* in Southeast China and Northwest India, respectively (Daniels and Roach 1987). This hypothesis is confirmed by diversity studies based on DNA analysis (D'Hont et al. 2008). Roach (1972) reports the range of *S. edule* to include Fiji, New Guinea, Indonesia and Malaysia.

Modern sugarcane cultivars produce multiple stems (culms) that can be several meters in length, each of which consists of alternating nodes and internodes. As the stem develops the leaves emerge, one leaf per node, attached at the base of the node, forming two alternate ranks on either side of the stem. The inflorescence, or tassle, of sugarcane is an open-branched panicle. The flowers consist of three stamens and a single carpel with a feathery stigma. The pollen grains are spherical when fertile and prismatic when sterile. Sugarcane fruit, called the caryopsis, is dry, indehiscent and one-seeded, and it cannot be separated from the seed. The sugarcane root system consists of adventitious and permanent shoot root types, that grow downwards for 5–7 meters (Cheavegatti-Gianotto et al. 2011; OGTR 2011).

History as a Biofuel Crop: Research Origin and Trend of Popularization

Bioethanol is a form of renewable energy that can be produced from agricultural feedstocks, and is produced through the fermentation of sugars or starch mainly from sugarcane, sweet sorghum, sugar beets and corn. Sugarcane ethanol is one of the most widely used biofuels, and Brazil is its largest producer. The Brazilian model, fine-tuned over decades, has low production costs, high energy ratio (production/consumption) and low greenhouse gas emissions when compared with other feedstock for biofuel. Ethanol from sugarcane as a biofuel in Brazil was motivated primarily by favourable climate conditions for this crop, and the Brazilian government's response to oil shocks through massive investment by the government in infrastructure and research.

In the 1930s, a Sugar and Alcohol Institute (IAA) was set up and it is responsible for establishing prices, production quotas of sugar mills and percentage blends in liquid fuel (blend gasoline with ethanol). In 1973, over 80% of the Brazilian petroleum was imported, and with the "First Oil Crisis" there was a significant increase in international oil prices. In response to high oil prices and the increasingly negative balance of trade, the government initiated major projects to change the energy matrix in Brazil. The "Second Oil Crisis" refers to oil prices increasing from US$24–26/barrel in 1979 to US$40 in 1981. At this point, ethanol became an important component of the Brazilian energy matrix, with increasing demand from the production of ethanol-only vehicles. The beginning of a new era of expansion with the technology of flex-fuel cars stimulated the internal market for alcohol as fuel.

This technology is based on sensors in the fuel system that automatically recognize the ethanol level in the fuel. The engine's electronic control unit then self-calibrates for the best possible operation; if ethanol is not present, the engine will self-calibrate to gasoline-only operation. The process is instantaneous and undetectable by the vehicle driver (Goldemberg 2008). Thus, consumers are free to choose which fuel they will use. Today in Brazil, the flex-fuel option is offered across the automobile industry, and cars based on both fuels have surpassed the number of cars working on gasoline alone. The Ethanol Program in Brazil is firmly established today, and is replacing approximately 40% of the gasoline that would otherwise be consumed in the country, at competitive prices (Goldemberg 2008). Presently all gasoline used in Brazil is blended with 20–25% anhydrous ethanol, a fuel with lower toxicity than fossil fuels.

Economic Importance

Sugarcane is an important crop for food and energy production. Worldwide cultivation of sugarcane occupies an area of 23.81 million hectare with a total production of 1,685 million metric tons (FAO STAT 2011). The main sugarcane-producing countries are Brazil (33%), India (23%), China (7%), Thailand (4%), Pakistan (4%), Mexico (3%), Colombia (3%), Australia (2%), the United States (2%) and the Philippines (2%). Out of 120 sugarcane producing countries, fifteen countries account for 86% of the total area, and they are responsible for 87.1% of global production. The average worldwide yield of sugarcane crops in 2010 was 70.7 tons per hectare, while the major producing countries are: India 66.7, China 65.7, Thailand 70, Pakistan 52.3, Mexico 71.6, Australia 77.6, United States 69.8 and Philippines 93.7 tons per hectare (FAO STAT 2011). Brazil is the world's largest sugarcane producer, with approximately 8.36 million cultivated hectares, which produced approximately 571 million tons in the 2011/2012 crop season (CONAB 2012). This is equivalent to a yield of 68.3 tons per hectare. Approximately half of the sugarcane is processed to produce 37 million tons of sugar (47.3%), and the remainder is used to produce 23 billion liters of ethanol (50.3%) (CONAB 2012). About 2.4% of sugarcane production is intended for animal feed, sugarcane juice, muscovado sugar, sugarcane syrup, rapadura, cachaca (type of brandy), and biopolymers, among other by-products. Sugarcane cultivation is concentrated in the south-central region of Brazil, which is responsible for approximately 70% of the national sugarcane production.

The nutritional composition of sugarcane is discussed in detail in OECD (2011). Sugarcane is a semi-perennial crop, therefore the plant is in the field all year round and the biomass could be subject to seasonal variation in its nutrient composition. The most important constituent in sugarcane is sucrose, which is typically measured in the plant stalk. Sucrose content can be quite variable (Berding 1997; Inman-Bamber et al. 2009). The composition of mature whole sugarcane is 67.5–80.9 moisture (% fresh weight), 1.8–4.1 crude protein (% DM = Dry matter), 22.7–35.9 crude fiber (% DM), 0.8–1.3 fat (% DM) and 1.2–6.2 ash (% DM). The chemical composition of sugarcane juice is mainly a high water content (about 85%) and contains 77–90% of total sugars (% DM), mainly sucrose and reducing sugars such as glucose and fructose. Sugarcane juice also contains amino acids with the most abundant ones being aspartic acid, glutamic acid, alanine and valine. Sugarcane bagasse typically contains approximately 40–50% moisture, and 43–58.5 crude fiber (% DM). The fiber fraction includes cellulose (35.8–58.4% DM), hemicellulose (16.4–42.4% DM) and lignin (9–22.3% DM).

The traditional purpose of sugarcane cultivation has been to produce sugar for human consumption. In recent years, economic interest in sugarcane has increased significantly due to the increased worldwide demand for sustainable energy production, like biofuels. Ethanol can be produced by fermenting sucrose from the cane; either that contained in the cane juice, as is the common model in Brazil, or the residual fermentable sugars contained in molasses, a by-product of sugar manufacture, which is the case in Australia (Renouf et al. 2011). The remaining fiber (bagasse) from the crushed cane is commonly combusted in mill boilers to produce steam and electricity for use in the sugar mill. If the bagasse is used efficiently, it is possible that sugar mills generate excess electric energy. This surplus electricity can be used for providing electrical energy to nearby cities. In future, the bagasse may also be used as raw material for ethanol production through acid or enzymatic hydrolysis, where cellulose and hemicellulose fractions can be converted into hexoses and pentoses. However, this technology is still under development, and its economic feasibility has yet to be proven.

Bioplastics are becoming increasingly important in the context of green chemistry since they can be made from renewable resources and can degrade easily later in nature or under industrial conditions. Among all the raw materials that are available for biopolymer production, sugarcane has one of the best profiles as a carbon source. Cane sugar, sucrose, is considered a versatile raw material for the fermentation industry to yield green chemicals. The most important biopolymers are polylactate (PLA) polyhydroxyalkanoate (PHA) and starch polymers (PA) (Bevan and Franssen 2006; Bengtsson et al. 2010; Wim and Tobias 2010).

Use as a Model Plant for Bioenergy Research

Bioenergy represents one of the best alternatives for capturing and storing solar energy through photosynthesis. This strategy is capable of meeting the urgent demands to reduce greenhouse gas emissions, improving the air quality in cities and replacing conventional energy sources such as petroleum-based fuels. There are a number of plant species that generate high yields of biomass with minimal inputs; many of these are C4 grasses. These plants possess an additional CO_2 concentrating mechanism that enables them to outperform C3 species, particularly under high temperature and light conditions and are, therefore, capable of generating larger quantities of biomass, even in resource limited environments. C4 grass species, such as miscanthus (*Miscanthus giganteus*), sweet sorghum (*Sorghum bicolor*), sugarcane and switchgrass (*Panicum virgatum*), are ideal energy crops because they possess the following traits: high conversion efficiency of light into biomass energy, high water use efficiency and high nitrogen use efficiency at the leaf level (Taylor et al. 2010; Somerville et al. 2010), capacity to grow in marginal land areas, and a relatively high tolerance to soil constraints such as salinity and waterlogging (Byrt et al. 2011).

Sugarcane has some advantage as a model plant for bioenergy. Sugarcane has a higher annual average productivity (80 MT ha^{-1} $year^{-1}$) as compared to other high-yield bioenergy crops such as miscanthus (15–40 MT ha^{-1} $year^{-1}$) and maize (10 MT ha^{-1} $year^{-1}$) (Somerville et al. 2010). Furthermore, experimental maximum yield in sugarcane is approximately 212 t/(ha year) fresh weight or 98 t/(ha year) dry biomass. However, this level of yield remains lower than the theoretical yield maxima that can reach cane yield of 381 t/(ha yr) or 177 t/(ha yr) dry biomass (Waclawovsky et al. 2010). Sugarcane crop has a high and positive energy balance, which indicates that the cultivation of this crop for biofuels can be economically and environmentally sustainable. Moreover, it has positive CO_2 balance and low net energy ratio (Yuan et al. 2008).

Although several feedstocks have good bioenergy potential, they are not well characterized agronomically, which makes it difficult for large scale cultivation. On the other hand, sugarcane crop management is well established, because there is a wide range of knowledge on the main pests and diseases related to this crop, harvesting and planting technologies, and a germplasm collection with broad genetic diversity for improvement on sugar or lignocellulosic traits. Additionally, perennial crops, like sugarcane, are more desirable than annuals as bioenergy feedstocks since they do not need to be reseeded each growing season and therefore cultivation costs are lower.

Overall Suitability as a Bioenergy Plant

While sugarcane is an efficient model as a bioenergy crop, some challenges must be overcome for its optimization. Sugarcane researchers should aim to integrate advances in biotechnology, genomic research, plant physiology knowledge and molecular marker applications with conventional breeding practices. This systems approach will help increase bioenergy production from this crop with reduced environmental impacts.

Further improvement of sugarcane through classical breeding programs has several challenges. First, the genetic base of sugarcane germplasms needs to be broadened with the aim of increasing the genetic variability for genotypes with high biomass and cold/drought tolerance, possibly through introgression of *Miscanthus* and *Erianthus* germplasm. Second, high-throughput phenotyping platforms need to be developed for use in the early stages of breeding or in the characterization of germplasm collections in order to accelerate the speed of selection. Furthermore, it is necessary to construct standardized protocols to phenotypically characterize and document germplasms for traits such as drought and flood tolerance, cold tolerance, photosynthetic efficiency (Lam et al. 2009).

To help achieve these objectives, genetic breeding programs should broaden the molecular marker studies in order to implement Marker Assisted Selection. The information from molecular markers will be crucial for genetic studies in sugarcane. Therefore, reliable linkage maps based on molecular markers are required to increase sequence assembly precision. New classes of molecular markers such as SNPs (Single Nucleotide Polymorphisms), statistical methods based on polyploid genetic and linkage disequilibrium analysis are important tools to make marker assisted selection in sugarcane a reality for breeding of sugarcane (Dal-Bianco et al. 2012). Lastly, the current physiological knowledge in sugarcane is far from ideal. Improved scientific understanding of the sugarcane plant, through the combination of traditional plant physiological studies and molecular techniques will be essential. Sucrose synthesis and translocation pathways, cell wall composition and pathways for lignin synthesis are examples of relevant biological problems that need to be more thoroughly understood (Lam et al. 2009; Soccol et al. 2010). It will also be necessary to develop bioinformatics tools to explore the increase in the amount of data related to sugarcane genomics and functional genomics. The assembly of a reference genome sequence for this crop is paramount to aid both the development of transgenics and the marker-assisted improvement of this crop. On the other hand, sugarcane transformation methods have yet to be improved. While this technique is feasible, sugarcane transformation remains hindered by low transformation efficiency, transgene inactivation and somaclonal variation issues (Dal-Bianco et al. 2012).

References

Amalraj, V.A. and N. Balasundaram. 2006. On the taxonomy of the members of '*Saccharum* complex'. Genetics Resources and Crop Evolution 53: 35–41.

Bakker, H. 1999. Sugarcane cultivation and management. Kluwer Academic/Plenum Publishers, New York.

Barnes, A.C. 1974. The sugarcane (2nd edition). Leonard Hill Books, London.

Bengtsson, S., A.R. Pisco, M.A.M. Reis and P.C. Lemos. 2010. Production of polyhydroxyalkanoates from fermented sugar cane molasses by a mixed culture enriched in glycogen accumulating organisms. Journal of Biotechnology 145: 253–263.

Berding, N. 1997. Clonal Improvement of Sugarcane Based on Selection for Moisture Content: Fact or Fiction. pp. 245–253. *In*: B.T. Egan (ed.). Proceedings of the 1997 Conference of the Australian Society of Sugar Cane Technologists. Watson Ferguson and Company, Brisbane.

Besse, P., C.L. McIntyre and N. Berding. 1997. Characterisation of *Erianthus* sect. *Ripidium* and *Saccharum* germplasm (Andropogoneae-Saccharinae) using RFLP markers. Euphytica 93: 283–292.

Bevan, M.W. and M.C.R. Franssen. 2006. Investing in green and white biotech. Nature Biotechnology 24: 765–767.

Bhat, N.R., K.S. Subba Rao and P.A. Kandasami. 1962. A review of cytogenetics of *saccharum*. Indian Journal Sugarcane Resources 7: 1–16.

Brandes, E.W., G.B. Sartoris and C.O. Grassl. 1938. Assembling and evaluating wild forms of sugarcane and closely related plants. Proceedings of ISSCT 6: 128–151.

Brandes, E.W. 1958. Origin, classification and characteristics. *In*: E. Artschwager and E.W. Brandes (eds.). Sugarcane (*Saccharum officinarum* L.) U.S. Department of Agriculture Handbook 122: 1–35.

Brown, J.S., R.J. Schnell, E.J. Power, S.L. Douglas and D.N. Kuhn. 2007. Analysis of clonal germplasm from five *Saccharum* species: *S. barberi*, *S. robustum*, *S. officinarum*, *S. sinense* and *S. spontaneum*. A study of inter- and intra species relationships using microsatellite markers. Genetics Resources Crop Evolution 54: 627–648.

Buzacott, J.H. 1965. Cane varieties and breeding. pp. 220–253. *In*: N.J. Kim, R.W. Mungomery and C.G. Hughes (eds.). Manual of Cane Growing. Sydney: Angus and Robertson.

Byrt, C.S., C.P.L. Grof and R.T. Furbank. 2011. C4 plants as biofuel feedstocks: Optimising biomass production and feedstock quality from a lignocellulosic perspective. Journal Integrative Plant Biology 53: 120–135.

Cheavegatti-Gianotto, A., H.M.C. Abreu, P. Arruda, Filho J.C. Bespalhok, W.L. Burnquist, S. Creste, L. Di Ciero, J.A. Ferro, A.V.O. Figuerira, T.S. Filgueira, M.F. Grossi-de-Sá, E.C. Guzzo, H.P. Hoffmann, M.G.A. Landell, N. Macedo, S. Matsuoka, F.C. Reinach, E. Romano, W.J. Silva, Filho M.C. Silva and E.C. Ulian. 2011. Sugarcane (*Saccharum X officinarum*): A reference study for the regulation of genetically modified cultivars in Brazil. Tropical Plant Biology 4: 62–89.

CONAB. 2012. Acompanhamento da Safra Brasileira. http://www. conab.gov.br/conabweb/ download/safra/dez2011.pdf. Cited 27 Jan 2012.

Cox, M., M. Hogarth and G. Smith. 2000. Cane Breeding and Improvement. pp. 91–108. *In*: M. Hogarth and P. Allsopp (eds.). Manual of Cane growing, Brisbane: Bureau of Sugar Experiment Stations.

D'Hont, A., L. Grivet, P. Feldmann, S. Rao, N. Berding and J.C. Glaszmann. 1996. Characterisation of the double genome structure of modern sugarcane cultivars (*Saccharum* spp.) by molecular cytogenetics. Molecular Genes Genetics 250: 405–413.

D'Hont, A., D. Ison, K. Alix, C. Roux and J.C. Glaszmann. 1998. Determination of basic chromosome numbers in the genus *Saccharum* by physical mapping of ribosomal RNA genes. Genome 41: 221–225.

D'Hont, A. and M. Layssac. 1998. Analysis of cultivars genome structure by molecular cytogenetics and the study of introgression mechanisms. France: Centre de coope´ration internationale en recherche agronomique pour le de´veloppement, Annual Crops Department (CIRAD-CA), 9–10.

D'Hont, A., F. Paulet and J.C. Glaszmann. 2002. Oligoclonal interspecific origin of 'North Indian' and 'Chinese' sugarcanes. Chromosome Research 10: 253–262.

D'Hont, A., G.M. Souza, M. Menossi, M. Vincentz, M.A. van-Sluys, J.C. Glaszmann and E. Ulian. 2008. Sugarcane: a major source of sweetness, alcohol, and bio-energy. *In*: P.H. Moore and R. Ming (eds.). Plant genetics and genomics: crops and models. New York: Springer 1: 483–513.

Dal-Bianco, M., M.S. Carneiro, C.T. Hotta, R.G. Chapola, H.P. Hoffmann, A.A. Garcia and G.M. Souza. 2012. Sugarcane improvement: how far can we go? Current Opinion in Biotechnology 23: 265–270.

Daniels, J. and B.T. Roach. 1987. Taxonomy and evolution. *In*: D.J. Heinz (ed.). Sugarcane Improvement through Breeding. Amsterdam: Elsevier 1: 7–84.

FAO STAT. 2011. Crops database: <http://faostat3.fao.org/home/index.html>. Accessed December 2011.

Galloway, J.H. 1989. The Sugar Cane Industry: An Historical Geography from its Origins to 1914. Cambridge Univ. Press, London.

Goldemberg, J. 2008. The Brazilian biofuels industry. Biotechnology Biofuels 1: 1–6.

Grassl, C.O. 1977. The origin of sugar producing cultivars of *Saccharum*. Sugarcane Breeding Newsletter 39: 8–33.

Grassl, C.O. 1967. Introgression between *Saccharum* and *Miscanthus* in N. Guinea & Pacific. Proceedings of ISSCT 12: 995–1003.

Grivet, L., C. Daniels, J.C. Glaszmann and A. D'Hont. 2004. A review of recent molecular genetics evidence for sugarcane evolution and domestication. Ethnobotany Research Application 2: 9–17.

Hemaprabha, G. and S.R. Sree Rangasamy. 2001. Genetic similarity among five species of *Saccharum* based on isozyme and RAPD markers. Indian Journal Genetics 61: 341–347.

Inman-Bamber, N.G., G.D. Bonnett, M.F. Spillman, M.L. Hewitt and J. Xu. 2009. Source-sink differences in genotypes and water regimes influencing sucrose accumulation in sugarcane stalks. Crop & Pasture Science 60: 316–327.

Irvine, J.E. 1999. *Saccharum* species as horticultural classes. Theoretical Applied Genetics 98: 186–194.

Lakshmanan, P., R.J. Geijskes, K.S. Aitken, C.L.P. Grof, G.D. Bonnett and G.R. Smith. 2005. Sugarcane Biotechnology: the Challenges and Opportunities. *In Vitro* Cell Developmental Biology-Plant 41: 345–363.

Lam, E., J. Shine, Jr., J. Da Silva, M. Lawton, S. Bonos, M. Calvino, H. Carrer, M.C. Silva-Filho, N. Glynn, Z. Helsel, J. Ma, E. Richard, Jr., G.M. Souza and R. Ming. 2009. Improving sugarcane for biofuel: engineering for an even better feedstock. Global Change Biology Bioenergy 1: 251–255.

Mohan Naidu, K. and T.V. Sreenivasan. 1987. Conservation of sugarcane germoplasm. pp. 33–53. *In*: Copersucar International Sugarcane Breeding Workshop, São Paulo: Copersucar.

Mukherjee, S.K. 1957. Origin and distribution of *Saccharum*. Botanical Gazette 119: 55–61.

OECD (Organisation for Economic Cooperation and Development). 2011. "Consensus document on compositional considerations for new varieties of sugarcane (*Saccharum* ssp. hybrids): Key food and feed nutrients, anti-nutrients and toxicants", Series on the Safety of Novel Foods and Feeds No.23, OECD Environment Directorate, Paris. Available online on the Biotrack website at http://www.oecd.org/science/biotrack/48962816.pdf (accessed May 2015).

OGTR. 2011. The biology of the *Saccharum* spp. (Sugarcane). http://www.health.gov.au/internet/ogtr/publishing.nsf/Content/sugarcane-3/$FILE/biologysugarcane08.pdf. Cited 12 Jan 2012.

Panje, R.R. and C.N. Babu. 1960. Studies in *S. spontaneum* distribution and geographical association of chromosome numbers. Cytologia 25: 152–172.

Parthasarathy, N. 1946. The probable origin of north Indian sugarcanes. Journal Indian Botanical Society, 133–150.

Price, S. 1965. Interspecific hybridization in sugarcane breeding. Proceedings of ISSCT 12: 1021–1026.

Renouf, M.A., R.J. Pagan and M.K. Wegener. 2011. Life cycle assessment of Australian sugarcane products with a focus on cane processing. International Journal Life Cycle Assessment 16: 125–137.

Roach, B.T. 1972. Chromosome numbers in *Saccharum edule*. Cytologia 37: 155–161.

Roach, B.T. and J. Daniels. 1987. A review of the origin and improvement of sugarcane. *In*: Copersucar International Sugarcane breeding Workshop 1: 1–31.

Soccol, C.R., L.P.S. Vandenberghe, A.B.P. Medeiros, S.G. Karp, M.S. Buckeridge, L.P. Ramos, A.P. Pitarelo, V. Ferreira-Leitão, L.M.F. Gottschalk and M.A. Ferrara. 2010. Bioethanol from lignocelluloses: Status and perspectives in Brazil. Bioresource Technology 101: 4820–4825.

Somerville, C., H. Youngs, C. Taylor, S.C. Davis and S.P. Long. 2010. Feedstocks for lignocellulosic biofuels. Science 329: 790–792.

Sreenivasan, T.V., B.S. Ahloowalia and D.J. Heinz. 1987. Cytogenetics. pp. 211–254. *In*: D.J. Heinz (ed.). Sugarcane improvement through breeding. Elsevier, Amsterdam.

Tai, P.Y.P. and J.D. Miller. 2001. A core collection for *Saccharum spontaneum* L. from the world collection of sugarcane. Crop Science 41: 879–885.

Taylor, S.H., S.P. Hulme, M. Rees, B.S. Ripley, F.I. Woodward and C.P. Osborne. 2010. Ecophysiological traits in C3 and C4 grasses: a phylogenetically controlled screening experiment. New Phytologist 185: 780–791.

Waclawovsky, A.J., P.M. Sato, C.G. Lembke, P.H. Moore and G.M. Souza. 2010. Sugarcane for bioenergy production: an assessment of yield and regulation of sucrose content. Plant Biotechnology Journal 8: 1–14.

Wim, J.G. and B. Tobias. 2010. Life cycle assessment of the manufacture of lactide and PLA biopolymers from sugarcane in Thailand. International Journal Life Cycle Assessment 15: 970–984.

Yuan, J.S., K.H. Tiller, H. Al-Ahmad, N.R. Stewart and C.N. Stewart, Jr. 2008. Plants to power: bioenergy to fuel the future. Trends in Plant Science 13: 421–429.

CHAPTER 2

Hybridization of Sugarcane and Other Grasses for Novel Traits

Jong-Won Park and *Jorge A. da Silva**

ABSTRACT

Modern commercial sugarcane cultivars are derived from the interspecific hybridization between *S. officinarum* and *S. spontaneum*. The high degree of polyploidy and the narrow gene pool of these cultivars impose difficulties on the effort of sugarcane breeders for genetic improvement. Fortunately, unlike other candidates for biofuel feedstock crops, *Saccharum* species are amenable to generate intergeneric hybrids with many other wild grass species by conventional breeding approaches. For this reason, currently many sugarcane breeding programs around the world are trying to incorporate more diverse germplasms, including sorghum and other closely related wild grass species, into the genetic background of modern sugarcane cultivars in an attempt to increase the genetic diversity. However, despite the lenience of *Saccharum* species for wide hybridization, only limited successful cases were reported due to the presence of strict reproductive barriers between individual plant species. Like self-incompatibility preventing plant inbreeding, a similar mechanism responsible for cross-incompatibility takes place

Texas A&M AgriLife Research, The Texas A&M University System, 2415 East Highway 83, Weslaco, TX 78596, USA.
* Corresponding author

during wide hybridization. The reproductive barriers mediating cross-incompatibility can be divided into two groups, pre- and post-zygotic barriers. Our work with the sorghum *iap* (inhibition of alien pollen) line for wide hybridization with *Saccharum* spp. showed that pre-zygotic barriers can be avoided, as well as post-zygotic barriers, by 2n gamete formation, common in *Saccharum* species. *Miscanthus* species is also useful for wide hybridization with *Saccharrum* species. Among the attributes to be introgressed from *Miscanthus*, cold and drought tolerance can be the most important ones for a low-input biofuel crop. However, the most probable source of valuable genetic traits for stress resistance is *S. spontaneum*, a highly polymorphic species, which we have been using in the past few years in hybridization crosses with sugarcane. The capability of wide hybridization of *Saccharum* species with other wild species makes sugarcane superior to most of the other candidates for biomass feedstock crops.

The Need for Hybridization of Sugarcane with Other Grasses and its Potential Benefits

Sugarcane (*Saccharum* spp. L.) is a perennial monocotyledonous crop widely cultivated in tropical and subtropical regions around the world as a major source for raw sugar. It is estimated that 70% of the global sugar production is from sugarcane and the rest is from sugar beet cultivated in temperate regions (Manners 2011). In addition, sugarcane has long been used for ethanol production in Brazil by the fermentation of sugarcane juice (Lam et al. 2009; Waclawovsky et al. 2010). Due to the limited amount of fossil fuel left in nature, there is an increasing consensus among scientists and governments to incorporate lignocellulosic biomass as a feedstock for biofuel production. Although the idea of using biomass for biofuel production has long been a subject among scientists (Lipinsky 1978; Grassl 1980; Clark et al. 1981), it now seems more economically feasible than ever before.

Lignocellulosic biomass is the most abundant and sustainable feedstock in nature which has an estimated annual production of about 200 billion tones (Sierra et al. 2008; Zhang 2008). At present, global sugarcane cultivation provides the largest scale of lignocellulosic biomass than any other crop species including the potential bioenergy feedstock crops, *Miscanthus* and switchgrass (Heaton et al. 2008; Waclawovsky et al. 2010; Manners 2011). Despite all the superior traits of sugarcane as a lignocellulosic biomass feedstock crop, adopting commercial sugarcane hybrids as a dedicated biomass crop for biofuel production is limited only to certain parts of the world, due to its lack of cold and drought stress tolerance. To expand the areas of sugarcane cultivation, it is necessary to enhance its tolerance to abiotic stress, with further studies at the genomic and molecular level to assist sugarcane breeding.

Modern commercial sugarcane cultivars (2n = 100–130) are derived from the interspecific hybridization between *S. officinarum* (2n = 80), high in sucrose, and *S. spontaneum* (2n = 40–128), tolerant to a broad range of biotic and abiotic stresses, followed by nobilization through a series of backcrossing of the progenies with other *S. officinarum* accessions (Stevenson 1965; Berding and Roach 1987). Although this interspecific hybridization contributed significantly to establish modern sugarcane cultivars, the high degree of polyploidy and the narrow gene pool of modern cultivars imposed difficulties on the effort of sugarcane breeders to develop new sugarcane cultivars with higher yields or enhanced disease resistance. In fact, it is believed that the increase in sugar yield through traditional sugarcane breeding has reached a stationary phase, mainly due to the narrow gene pool available for sugarcane breeding (Mariotti 2002; Dillon et al. 2007). Currently, many sugarcane breeding programs around the world are trying to incorporate more diverse germplasms, including sorghum and other closely related wild grass species, into the genetic background of modern sugarcane cultivars in an attempt to increase the genetic diversity (Dillon et al. 2007; Hodnett et al. 2010).

Mukherjee (1957) used the term '*Saccharum* complex' to describe the genus *Saccharum* and three closely related genera: *Erianthus* (sect. *Ripidium*) (2n = 20–60), *Sclerostachya* (2n = 30) and *Narenga* (2n = 30). Later, Daniels et al. (1975) included the genus *Miscanthus* (sect. *Diandra*) (2n = 38–76) in the *Saccharum* complex (Sreenivasan et al. 1987). The members of the *Saccharum* complex, especially the genera *Erianthus* and *Miscanthus*, are thought to have contributed to the formation of genus *Saccharum* over a long period of time (Daniels and Roach 1987; Grivet et al. 2006). Chloroplast genome analysis showed that the genera within the *Saccharum* complex are closely related to each other (Sobral et al. 1994; Aitken and McNeil 2010). According to DNA sequence analyses of the ITS (Internal Transcribed Spacer) region of nuclear ribosomal DNA (Hodkinson et al. 2002), the genus *Saccharum* is considered more closely related to the genus *Miscanthus* than any other members of the *Saccharum* complex (Aitken and McNeil 2010).

Since the successful interspecific hybridization between *S. officinarum* and *S. spontaneum* in the early 1900s (Ming et al. 2010), the members of the *Saccharum* complex have been considered an excellent genetic resource for sugarcane improvement through wide hybridization (Price 1965; D'Hont et al. 1995; Besse et al. 1997; Piperidis et al. 2000). In fact, sugarcane breeders have tried wide hybridization with members of the *Saccharum* complex in an attempt to increase genetic diversity as well as to enhance the biotic and abiotic stress tolerance of sugarcane, but with limited success (Aitken and McNeil 2010).

The Obstacles of Creating Sugarcane Hybrids

Since diverse wild species and land races provide useful gene pools for crop improvement, interspecific wide hybridization has been attempted frequently in order to broaden genetic diversity of crop species as well as to transfer favorable genetic traits from wild species into major crops (Ishikawa and Kinoshita 2009). However, in spite of numerous efforts for interspecific wide hybridization, little progress has been made, mainly due to the presence of strict reproductive barriers between individual plant species (Bates and Deyoe 1973; Baum et al. 1992; Dresselhaus et al. 2011). As shown in self-incompatibility, which prevents inbreeding in many plant species (Lundqvist 1956; Martinez-Reyna and Vogel 2002), a similar mechanism responsible for cross-incompatibility takes place during wide hybridization (Baum et al. 1992; Dresselhaus et al. 2011). Therefore, it is essential to study and adopt a strategy to overcome the existing reproductive barriers for successful wide hybridization.

The reproductive barriers mediating cross-incompatibility are divided into two groups, pre- and post-zygotic barriers (Baum et al. 1992; Dresselhaus et al. 2011). Numerous reports have shown pre-zygotic barriers to be the main cause of unsuccessful interspecific wide hybridization in a variety of plant species including *Sorghum* (Hodnett et al. 2005), *Oryza* (Fu et al. 2007), *Gossypium* (Ram et al. 2008), *Solanum* (Hermsen and Ramanna 1976), *Pennisetum* (Kaushal and Sidhu 2000), *Panicum* (Martinez-Reyna and Vogel 2002) and *Jatropha* spp. (Kumar et al. 2009). Pre-zygotic barriers also determine the incompatibility between pollen and pistil before fertilization (Baum et al. 1992). The outcome of the incompatible interaction at the pre-zygotic level includes failure of pollen germination, arrest of pollen tube growth through stigma and style, followed by swelling or rupture of pollen tube tip, all of which lead to the failure of fertilization (Baum et al. 1992; Dresselhaus et al. 2011).

Although the pre-zygotic barriers provide a highly specialized cross-incompatibility system, a study with the sorghum *iap* (*inhibition of alien pollen*) line for wide hybridization with *Saccharum* spp. showed that pre-zygotic barriers can be avoided (Hodnett et al. 2010). Although the sorghum *iap* line greatly facilitated the hybrid generation between sorghum and sugarcane, the success rate of sorghum *iap* x sugarcane wide hybridization was affected by sugarcane genotype, suggesting the presence of genotype-dependent compatibility (Hodnett et al. 2010). Recently, Alarmelu and Shanthi (2011) showed the inter-varietal incompatibility in sugarcane. They showed that the pre-zygotic barriers are responsible for the interrupted inter-varietal hybridization between several self-incompatible commercial sugarcane hybrids (Alarmelu and Shanthi 2011). To our knowledge, this study was the first report in the *Saccharum* species showing evidence of pre-zygotic incompatibility between sugarcane hybrids, which may have an effect on sugarcane wide hybridization as well. It is not clear at the moment whether the genotype-specific compatibility observed in the sorghum *iap* line x sugarcane wide hybridization is due to the pre-zygotic barriers responsible for inter-varietal incompatibility in sugarcane or due to post-zygotic barriers after fertilization.

Post-zygotic incompatible reaction is usually related to the interploidy and interspecific hybridization, resulting in embryo death due to abnormal endosperm development (Johnston et al. 1980), production of sterile hybrids caused by nonviable gamete formation and potential chromosome elimination of one parental line during embryo development as seen in the cross between wheat and maize (Laurie and Bennett 1988; Baum et al. 1992). During the double fertilization process in angiosperms, one sperm cell (n) fertilizes an egg cell (n) to produce a diploid embryo and another sperm cell fuses with a central cell (2n) in the female gametophyte to initiate a triploid (3n) endosperm development which supplies vital developmental nutrients to the embryo (Liu et al. 2010). The abnormal development of endosperm is directly related to the death of a viable hybrid embryo commonly observed in interploidy wide hybridization (Ishikawa and Kinoshita 2009). A similar phenomenon has been observed in sugarcane wide hybridization with sorghum where very few viable hybrids were produced (Nair 1999). Although tissue culture method can rescue the viable embryo produced by wide hybridization (Rodrangboon et al. 2002; Price et al. 2005), the whole process for a successful embryo rescue by tissue culture needs to be optimized for each plant species (Rodrangboon et al. 2002) and can cause unpredictable somaclonal variation on the rescued embryos (Larkin and Scowcroft 1981).

Johnston et al. (1980) pointed out the importance of the genic ratio between maternal and paternal parents in the endosperm after fertilization. They proposed an endosperm balance number (EBN) hypothesis where 2:1 maternal to paternal EBN ratio in the triploid endosperm is essential to ensure the successful endosperm development during wide hybridization (Johnston et al. 1980). In the hypothesis, EBN does not exactly correspond to the ploidy level of a plant genome, but instead, EBN is empirically assigned to each plant species based on its crossing behavior to a standard species (Johnston et al. 1980). Similar hypotheses, 'Polar nuclei activation' and 'Ploidy barrier to endosperm', were proposed from studies in oat (Nishiyama and Yabuno 1978) and maize (Lin 1984), respectively.

In 1993, Burner and Legendre (1993) generated wide hybrids from crosses between commercial sugarcane hybrids (female) and *S. spontaneum* (male) from which they observed a 'n+n' chromosome transmission in the hybrids. They implicated that sugarcane hybrids and *S. spontaneum* may share a common EBN that ensured the 2:1 EBN ratio in the triploid endosperm (Burner and Legendre 1993). Although EBN is not exactly related to the ploidy level of a plant genome, Johnston et al. (1980) suggested that in certain cases, the unreduced 2n gamete formation in female gametophyte provides a way to overcome the post-zygotic barrier imposed by a EBN ratio deviated from 2:1 maternal to paternal EBN ratio (Johnston et al. 1980). According to Johnston et al. (1980), it may be possible that the 2n gamete formation during the interspecific hybridization between *S. officinarum* and *S. spontaneum* established in the early 1900s may have served as a base to overcome the obstacle of non 2:1 EBN ratio between *S. officinarum* (female) and *S. spontaneum* (male) (Roach 1972; Johnston et al. 1980). However, further study is needed to understand the relevance of the EBN hypothesis and 2n gamete involvement on sugarcane wide hybridization.

Another outcome of post-zygotic incompatibility is the production of sterile hybrids due to aberrant meiosis. This could be caused by a lack of chromosome pairing and leads to non-viable gamete formation (Fu et al. 2007). A lack of fertility among hybrids imposes a great hurdle for further incorporation of beneficial genetic traits into crop species by backcrossing. In sugarcane, the hybrid sterility as well as genome elimination were a common problem in wide hybridization with *E. arundinaceus* (D'Hont et al. 1995; Besse et al. 1997; Piperidis et al. 2000). However, Deng et al. (2002) successfully generated fertile F_1 hybrids from *S. officinarum* x *E. arundinaceus*, which were used to generate BC_1 progenies by a cross between the F_1 and sugarcane hybrids. According to genome *in situ* hybridization (GISH) studies (Piperidis et al. 2010), BC_1 hybrids were derived from 2n+n chromosome transmission, suggesting that the sterility of F_1 hybrid progenies was overcome by the formation of unreduced 2n gametes in the F_1 hybrids (Piperidis et al. 2010).

2n gamete formation in sugarcane is not an unusual event as observed in the early *Saccharum* interspecific wide hybridization (Roach 1972). Since the production of unreduced 2n gametes is considered as a major driving force for the generation of polyploid plants (Bretagnolle and Thompson 1995; Ramsey and Schemske 1998), the induction of 2n gamete formation during wide hybridization can be a useful strategy to obtain more efficient wide hybridization for crop improvement. It has been shown that 2n gamete production is affected by environmental factors such as temperature

(McHale 1983; Bretagnolle and Thompson 1995) and chemical (e.g., nitrous oxide) treatment (Barba-Gonzalez et al. 2006). Therefore, the introduction of a controlled 2n gamete inducing system in sugarcane wide hybridization may improve the efficiency possibly by overcoming the post-zygotic barriers corresponding to the EBN hypothesis (Johnston et al. 1980) as well as the hybrid sterility issue (Piperidis et al. 2000; Fu et al. 2007).

Examples of Successful Hybrids and their Characterization

The members of the tribe Andropogoneae in the family Poaceae have an efficient C4 photosynthesis system, and it has been shown that many of them are crossable with *Saccharum* species (Sreenivasan et al. 1987). The details of sugarcane wide hybridization among the members of the *Saccharum* complex that took place in the early 20th century were well described by Sreenivasan et al. (1987). Only relatively recent sugarcane wide hybridization examples, of which hybrid status was confirmed either by cytological method or molecular markers, or both, are listed in Table 1. Among the members of the *Saccharum* complex, *Erianthus* species have been used most frequently for wide hybridization due to its good ratooning ability and tolerance to environmental stresses (D'Hont et al. 1995; Besse et al. 1997; Piperidis et al. 2000; Cai et al. 2005; Aitken et al. 2007; Aitken and McNeil 2010).

Another member of the *Saccharum* complex useful for wide hybridization with the *Saccharum* species is *Miscanthus*. *Miscanthus* is a perennial wild grass species growing in Asia and the Pacific Islands and is considered to be the closest relative of *Saccharum* species among the members of the *Saccharum* complex (Aitken and McNeil 2010). Frequent hybridization, including natural hybridization, between *Saccharum* and *Miscanthus* is considered to have taken place (Sobral et al. 1994; Hodkinson et al. 2002; Grivet et al. 2006; Manners 2011). Due to its wide geographical distribution ranging from temperate to tropical regions and from coastal to high altitude regions (Hodkinson et al. 2002), *Miscanthus* is of particular interest for sugarcane breeders as a source for cold and drought tolerance in commercial sugarcane hybrids. A cross between sugarcane and *M. sinensis* has been made to generate cold tolerant hybrid lines called Miscane (da Silva et al. 2011; Manners 2011; Park et al. 2011). Figure 1 shows commercial sugarcane (*Saccharum* spp.) and *M. sinensis* plants growing in tropical and temperate environment, respectively, and seedlings of Miscane along with seedlings of *M. sinensis* in the same row (green plants), surrounded by seedlings of sugarcane (with cold damage symptoms) growing in a field in Weslaco, Texas. The genome size analysis of some of these lines indicated

Table 1. List of Sugarcane Wide Hybridization.

Cross (female x male)	Wide hybrids	Hybridization status confirmed by	Characteristics	Reference
S. officinarum x *M. sinensis*	Yes	Cytology (somatic chromosome number)	Downy mildew resistance	(Chen et al. 1983)
S. officinarum x *E. arundinaceus*	Yes	Molecular marker; Genome *in situ* hybridization		(D'Hont et al. 1995)
S. officinarum x *E. arundinaceus*	Yes	Genomic slot blot hybridization		(Besse et al. 1997)
S. bicolor x *S. officinarum*	Yes	Cytology (somatic chromosome number)	Morphologically similar to *Saccharum*, but lacked vigor	(Nair 1999)
S. officinarum x *E. arundinaceus*	Yes	5S rDNA PCR; GISH	Sterile hybrids	(Piperidis et al. 2000)
S. officinarum x *E. arundinaceus*	Yes	Morphology	Cold tolerance and red rot resistance in hybrids	(Ram et al. 2001)
S. officinarum x *E. arundinaceus*	Yes	Isozyme marker; microsatellite marker; 5S rDNA PCR	Produced fertile F1 hybrids	(Deng et al. 2002; Cai et al. 2005)
Sugarcane x *S. bicolor*	Yes	RAPD marker	Somaclones of sugarcane x *S. bicolor* hybrids	(Nair et al. 2006)
S. officinarum x *Z. mays*	Yes	RAPD marker	Somaclones of *S. officinarum* x *Z. mays* hybrids	(Nair et al. 2006)
S. bicolor x *Saccharum* spp.	Yes	Morphology and cytology	Wide hybridization done with *S. bicolor* iap mutant line; seed set production affected by male parent's genetic/genomic compatibility	(Hodnett et al. 2010)

Figure 1. Pictures of commercial sugarcane (A) growing in tropical environment, (B) *M. sinensis* growing in temperate environment, and (C) seedlings of Miscane and *M. sinensis* on the same row (◀), surrounded by commercial sugarcane seedlings. The images in (C) was taken after an unexpected freeze during 2004 in Weslaco, Texas. The inset in (C) shows the difference in phenotypes on the leaves of sugarcane (top), Miscane (middle), and *M. sinensis* (bottom).

that they may maintain small amounts of sugarcane genome content, if any (unpublished data). Previous work involving wide hybridization of these two species has shown that the introgression of *Miscanthus* germplasm into sugarcane resulted in the resistance against downy mildew (Chen et al. 1983).

In addition to the members of the *Saccharum* complex, sorghum has also been used for wide hybridization with sugarcane by using a sorghum *iap* line to facilitate wide hybridization between sorghum and sugarcane (Hodnett et al. 2010). Since sorghum shows superior drought tolerance compared to sugarcane, the hybrid lines will be a useful gene pool for the development of sugarcane drought tolerance (da Silva et al. 2011; Manners 2011).

Future Perspectives on the Application of Wide Hybridization to Extend the Range and Productivity of Sugarcane

Since the two major achievements in the history of sugarcane breeding, (1) the discovery of sugarcane fertility and (2) the incorporation of *S. spontaneum* germplasm into *S. officinarum* that resulted in high vigor and disease resistance in the hybrids (Stevenson 1965; Ming et al. 2010), many sugarcane breeding programs around the world have looked into wild species as a source for genetic improvement of sugarcane through wide hybridization (Aitken and McNeil 2010). Being a member of the closely related inter-breeding group, the *Saccharum* complex, sugarcane can generate interspecific wide hybrids, combining important attributes as a feedstock for energy. Among these attributes, cold and drought tolerance can be the most important ones for a low-input biofuel crop. An example of a good genetic resource for sugarcane improvement is *Miscanthus* spp. *Miscanthus* spp. can be found from Tahiti to eastern Indonesia, Indochina to northern China, Siberia and Japan. In terms of altitude, this genus is found from sea level to 3,300 m above sea level, containing valuable genes for cold resistance (Lo et al. 1977).

Both sugarcane and sorghum have been identified as bioenergy crops. The hybridization of these bioenergy crops can allow the combination of the high biomass production of sugarcane with the drought tolerance and wide adaptation of sorghum. In addition, the ability of sorghum to be propagated by true seed would be a tremendous advantage to the sorghum-sugarcane wide hybrid, contrary to the current labor-intensive billet planting method used for sugarcane (Hodnett et al. 2010).

The most probable source of valuable positive genes for stress resistance, within the *Saccharum* complex, would be *S. spontaneum*, a highly polymorphic species growing in the tropics and subtropics, 8° S to 40° N (Brandes et al. 1939). The habitat of *S. spontaneum* extends from Africa and the Mediterranean through the Indian subcontinent to Japan and Indonesia/ New Guinea (Panje and Babu 1960). Being highly adaptable, *S. spontaneum* can grow in waterlogged conditions of marshes, with drought stress in the desert, and in saline conditions near the sea (Mukherjee 1950). *S. spontaneum* is also extremely resistant to a wide range of temperatures, from winter snow to tropical heat and can be found at sea level up to 2,700 m above sea level (Mukherjee 1950). Given its ability to propagate both through true seed, as well as vegetatively by lateral buds and rhizomes, *S. spontaneum* is considered a noxious weed in mainland US. Its good ratooning ability will allow its production over a number of years without having to re-plant, which is economically important given the high costs of crop re-planting.

Utilization of plant biomass as a feedstock for second-generation biofuel production has been a major issue among plant biologists, especially among plant breeders who try to develop new optimized biofuel feedstock crops (Somerville et al. 2010). Both *Miscanthus* (Heaton et al. 2010) and sorghum (Paterson et al. 2009) have been extensively investigated as biofuel feedstock crops together with switchgrass (Adler et al. 2006). Although sugarcane has long been used for mass production of first-generation ethanol by fermenting sugarcane juice, its potential as a renewable biomass feedstock crop for second-generation biofuel production has recently been realized due to its superior energy conversion efficiency for biomass production (Lam et al. 2009; Waclawovsky et al. 2010). However, its lack of cold and drought tolerance has limited its cultivation, mainly in tropical and subtropical regions. Fortunately, unlike other candidates for biofuel feedstock crops, *Saccharum* species are amenable to generate intergeneric hybrids with many other wild grass species by the conventional breeding process. This amenability of the *Saccharum* species is a particularly important aspect for generating an optimized biomass feedstock crop for the renewable biofuel production with little input.

As discussed in this review, a diverse array of reproductive barriers before and after fertilization imposes a significant obstacle for the generation of sugarcane wide hybrids. Furthermore, only limited information about sugarcane cross-incompatibility system is available. As a result, a more comprehensive research related to the reproductive barriers for sugarcane wide hybridization is mostly left unexploited. Recently, maize has been proposed as a model system to study cross-incompatibility in grasses (Dresselhaus et al. 2011). Since sugarcane and maize belong to the tribe

Andropogoneae in the family Poaceae (Dillon et al. 2007), the obtainable information from a maize cross-incompatibility study will help to understand the basis for a sugarcane cross-incompatibility system. At the same time, the photoperiod study for flower induction among the members of the *Saccharum* complex, and genomic and genetic compatibility studies between genotypes/species will greatly benefit the development of a better-suited, cane-based biomass feedstock crop. Recent advancement of the DNA sequencing technology opens an opportunity to overcome the extreme complexity of the sugarcane genome, allowing molecular marker development for marker-assisted selection of intergeneric hybrids. Considering that sugarcane is one of the most efficient biomass producers, the capability of wide hybridization of *Saccharum* species with other wild species makes sugarcane superior to most of the other candidates for biomass feedstock crops.

References

Adler, P.R., M.A. Sanderson, A.A. Boateng, P.I. Weimer and H.-J.G. Jung. 2006. Biomass yield and biofuel quality of switchgrass harvested in fall or spring. Agron. J. 98: 1518–1525.

Aitken, K. and M. McNeil. 2010. Diversity analysis. pp. 419–442. *In*: R. Henry and C. Kole (eds.). Genetics, Genomics and Breeding of Sugarcane. CRC Press, New York, USA.

Aitken, K.S., J. Li, L. Wang, C. Qing, Y.H. Fan and P.A. Jackson. 2007. Introgression of *Erianthus rockii* into sugarcane and verification of intergeneric hybrids using molecular markers. Genet. Resour. Crop Evol. 54: 1395–1405.

Alarmelu, S. and R.M. Shanthi. 2011. Incompatibility studies in sugarcane (*Saccharum* spp.). Indian J. Genet. Plant Breeding 71: 43–48.

Barba-Gonzalez, R., C.T. Miller, M.S. Ramanna and J.M. Van Tuyl. 2006. Nitrous oxide (N2O) induces 2n gamete in sterile F1 hybrids between Oriental x Asiatic lily (*Lilium*) hybrids and leads to intergenomic recombination. Euphytica 148: 303–309.

Bates, L.S. and C.W. Deyoe. 1973. Wide hybridization and cereal improvement. Econ. Bot. 27: 401–412.

Baum, M., E.S. Lagudah and R. Appels. 1992. Wide Crosses in Cereals. Annu. Rev. Plant Phys. 43: 117–143.

Berding, N. and B.T. Roach. 1987. Germplasm conservation, collection, maintenance and use. pp. 143–210. *In*: D.J. Heinz (ed.). Sugarcane improvement through breeding. Elsevier, Amsterdam.

Besse, P., C.L. McIntyre, D.M. Burner and C.G. de Almeida. 1997. Using genomic slot blot hybridization to assess intergeneric *Saccharum* x *Erianthus* hybrids (Andropogonae-Saccharinae). Genome 40: 428–432.

Brandes, E.W., G.B. Sartoris and C.O. Grassl. 1939. Assembling and evaluating wild forms of sugarcane and related plants. Proc. Int. Soc. Sugar Cane Technol. 6: 128–153.

Bretagnolle, F. and J.D. Thompson. 1995. Tansley Review No. 78. Gamete with the stomatic chromosome number: Mechanisms of their formation and role in the evolution of Autopolyploid plants. New Phytologist 129: 1–22.

Burner, D.M. and B.L. Legendre. 1993. Chromosome transmission and meiotic stability of sugarcane (*Saccharum* spp.) hybrid derivatives. Crop Science 33: 600–606.

Cai, Q., K. Aitken, H.H. Deng, X.W. Chen, C. Fu, P.A. Jackson and C.L. McIntyre. 2005. Verification of the introgression of *Erianthus arundinaceus* germplasm into sugarcane using molecular markers. Plant Breeding 124: 322–328.

Chen, H.W., Y.J. Huang, I.S. Shen and S.C. Shih. 1983. Utilization of *Miscanthus* germplasm in sugar cane breeding in Taiwan. Proc. Int. Soc. Sugar Cane Technol. 18: 641–649.

Clark, J.W., E.J. Soltes and F.R. Miller. 1981. Sorghum—A versatile, multi-purpose biomass crop. Biosour. Digest. 3: 36–57.

D'Hont, A., P.S. Rao, F. Feldmann, L. Grivet, N. Islam-Faridi, P. Taylor and J.C. Glaszmann. 1995. Identification and characterization of sugarcane intergeneric hybrids, *Saccharum officinarum* x *Erianthus arundinaceus*, with molecular markers and DNA *in situ* hybridisation. Theor. Appl. Genet. 91: 320–326.

da Silva, J.A., J.W. Park and Q. Yu. 2011. Sugarcane wide hybrids: New feedstocks for energy. 10th Germplasm and Breeding & 7th Molecular Biology Workshops for ISSCT, p. 21.

Daniels, J. and B.T. Roach. 1987. Taxonomy and evolution. pp. 7–84. *In*: D.J. Heinz (ed.). Sugarcane improvement through breeding, vol 11. Elsevier, Amsterdam.

Daniels, J., P. Smith, N. Paton and C.A. Williams. 1975. The origin of the genus *Saccharum*. Sugarcane Breed. Newslett. 36: 24–39.

Deng, H.H., Z.Z. Liao, Q.W. Li, F.Y. Loa, C. Fu, X.W. Chen, C.M. Zhang, S.M. Liu and Y.H. Yang. 2002. Breeding and isozyme marker assisted selection of F2 hybrids from *Saccharum* spp. x *Erianthus arundinaceus*. Sugarcane Canesugar 1: 1–5.

Dillon, S.L., F.M. Shapter, R.J. Henry, G. Cordeiro, L. Izquierdo and L.S. Lee. 2007. Domestication to crop improvement: genetic resources for *Sorghum* and *Saccharum* (Andropogoneae). Ann. Bot. 100: 975–989.

Dresselhaus, T., A. Lausser and M.L. Marton. 2011. Using maize as a model to study pollen tube growth and guidance, cross-incompatibility and sperm delivery in grasses. Ann. Bot. 108: 727–737.

Fu, X.L., Y.G. Lu, X.D. Liu, J.Q. Li and J.H. Feng. 2007. Cytological mechanisms of interspecific incrossability and hybrid sterility between *Oryza sativa* L. and *O. alta* Swallen. Chinese Sci. Bull. 52: 755–765.

Grassl, C.O. 1980. Breeding Andropogoneae at the generic level for biomass. Sugarcane Breed. Newslett. 43: 41–57.

Grivet, L., J.C. Glaszmann and A. D'Hont. 2006. Molecular evidences for sugarcane evolution and domestication. pp. 49–66. *In*: T. Motley et al. (eds.). Darwin's Harvest. New Approaches to the Orgins, Evolution, and Conservation of Crops. Columbia University Press, New York, USA.

Heaton, E.A., F.G. Dohleman and S.P. Long. 2008. Meeting US biofuel goals with less land: the potential of *Miscanthus*. Glob. Change Biol. 14: 2000–2014.

Heaton, E.A., F.G. Dohleman, A.F. Miguez, J.A. Juvik, V. Lozovaya, J. Widholm, O.A. Zabotina, G.F. McIsaac, M.B. David, T.B. Voigt, N.N. Boersma and S.P. Long. 2010. *Miscanthus*: A Promising Biomass Crop. Adv. Bot. Res. 56: 75–137.

Hermsen, J.G. and M.S. Ramanna. 1976. Barriers to hybridization of *Solanum bulbocastanum* DUN. and *S. verrucosum* SCHLECHTD. and structural hybridity in their F1 plants. Euphytica 25: 1–10.

Hodkinson, T.R., M.W. Chase, M.D. Lledo, N. Salamin and S.A. Renvoize. 2002. Phylogenetics of *Miscanthus*, *Saccharum* and related genera (Saccharinae, Andropogoneae, Poaceae) based on DNA sequences from ITS nuclear ribosomal DNA and plastid trnL intron and trnL-F intergenic spacers. J. Plant Res. 115: 381–392.

Hodnett, G.L., B.L. Burson, W.L. Rooney, S.L. Dillon and H.J. Price. 2005. Pollen-pistil interactions result in reproductive isolation between *Sorghum bicolor* and divergent *Sorghum* species. Crop Sci. 45: 1403–1409.

Hodnett, G.L., A.L. Hale, D.J. Packer, D.M. Stelly, J.A. da Silva and W.L. Rooney. 2010. Elimination of a reproductive barrier facilitates intergeneric hybridization of *Sorghum bicolor* and *Saccharum*. Crop Sci. 50: 1188–1195.

Ishikawa, R. and T. Kinoshita. 2009. Epigenetic Programming: The Challenge to Species Hybridization. Mol. Plant 2: 589–599.

Johnston, S.A., T.P.M. Dennijs, S.J. Peloquin and R.E. Hanneman. 1980. The significance of genic balance to endosperm development in interspecific crosses. Theor. Appl. Genet. 57: 5–9.

Kaushal, P. and J.S. Sidhu. 2000. Pre-fertilization incompatibility barriers to interspecific hybridizations in *Pennisetum* species. J. Agri. Sci. 134: 199–206.

Kumar, R.S., K.T. Parthiban, P. Hemalatha, T. Kalaiselvi and M.G. Rao. 2009. Investigation on cross-compability barriers in the biofuel crop *Jatropha curcas* L. with wild *Jatropha* species. Crop Sci. 49: 1667–1674.

Lam, E., J. Shine, Jr., J. Da Silva, M. Lawton, S. Bonos, M. Calvino, H. Carrer, M.C. Silva-Filho, N. Glynn, Z. Helsel, J. Ma, E. Richard, Jr., G.M. Souza and R. Ming. 2009. Improving sugarcane for biofuel: Engineering for an even better feedstock. GCB Bioenergy 1: 251–255.

Larkin, P.J. and W.R. Scowcroft. 1981. Somaclonal variation—A novel source of variability from cell cultures for plant improvement. Theor. Appl. Genet. 60: 197–214.

Laurie, D.A. and M.D. Bennett. 1988. The production of haploid wheat plants from wheat x maize crosses. Theor. Appl. Genet. 76: 393–397.

Lin, B.-Y. 1984. Ploidy barrier to endosperm development in maize. Genetics 107: 103–115.

Lipinsky, E.S. 1978. Fuels from biomass: Integration with food and materials systems. Science 1999: 644–651.

Liu, Y., Z. Yan, N. Chen, X. Di, J. Huang and G. Guo. 2010. Development and function of central cell in angiosperm female gametophyte. Genesis 48: 466–478.

Lo, C.C., Y.H. Chia, W.H. Chen, K.C. Shang, I.S. Shen and S.C. Shih. 1977. Collecting *Miscanthus* germplasm in Taiwan. Proc. Int. Soc. Sugar Cane Technol. 16: 59–68.

Lundqvist, A. 1956. Self-incompatibility in rye. I. Genetic control in the diploid. Hereditas 42: 293–348.

Manners, J.M. 2011. Functional genomics of sugarcane. Adv. Bot. Res. 60: 89–168.

Mariotti, J.A. 2002. Selection for sugar cane yield and quality components in subtropical climates. Sugar Cane Int. March/April: 22–26.

Martinez-Reyna, J.M. and K.P. Vogel. 2002. Incompatibility systems in switchgrass. Crop Sci. 42: 1800–1805.

McHale, N.A. 1983. Environmental induction of high frequency 2n pollen formation in diploid Solanum. Can J. Genet. Cyto. 25: 609–615.

Ming, R., P.H. Moore, K.K. Wu, A. D'Hont, J.C. Glaszmann, T.L. Tew, T.E. Mirkov, J.A. da Silva, J. Jifon, M. Rai, R.J. Schnell, S.M. Brumbley, P. Lakshmanan, J.C. Comstock and A.H. Paterson. 2010. Sugarcane improvement through breeding and biotechnology. pp. 15–118. *In*: J. Janick (ed.). Plant Breeding Reviews, vol. 27. John Wiley & Sons, Inc., Oxford, UK.

Mukherjee, S.K. 1950. Search for wild relatives of sugarcane in India. Proc. Int. Soc. Sugar Cane Technol. 52: 261–262.

Mukherjee, S.K. 1957. Origin and distribution of *Saccharum*. Bot. Gaz. 119: 55–61.

Nair, N.V. 1999. Production and cyto-morphological analysis of intergeneric hybrids of *Sorghum* x *Saccharum*. Euphytica. 108: 187–191.

Nair, N.V., A. Selvi, T.V. Sreenivasan, K.N. Pushpalatha and S. Mary. 2006. Characterization of intergeneric hybrids of *Saccharum* using molecular markers. Genet. Resour. Crop Evol. 53: 163–169.

Nishiyama, I. and T. Yabuno. 1978. Causal relationships between the polar nuclei in double fertilization and interspecific cross-incompatibility in *Avena*. Cytologia 43: 453–466.

Panje, R.R. and C.N. Babu. 1960. Studies in *Saccharum spontaneum*. Distribution and geographical association of chromosome numbers. Cytologia 25: 152–172.

Park, J.W., Q. Yu, N.S. Gracia, G.M. Acuna and J.A. da Silva. 2011. Development of new intergeneric hybrids, Miscanes, as a source of biomass feedstock for biofuel production. Proc. of the Plant and Animal Genomes Conference XIX: W319.

Paterson, A.H., J.E. Bowers, R. Bruggman, I. Dubchak, J. Grimwood, H. Gundlach, G. Haberer, U. Hellsten, T. Mitros, A. Pliakov, J. Schmutz, M. Spannagl, H. Tang, X. Wang, T. Wicker, A.K. Bharti, J. Chapman, F.A. Feltus, U. Gowik, I.V. Grigoriev, E. Lyons, C.A. Maher, M. Martis, A. Narechania, R.P. Otillar, B.W. Penning, A.A. Salamov, Y. Wang, L. Zhang, N.C. Carpita, M. Freeling, A.R. Gingle, T. Hash, B. Keller, P. Klein, S. Kresovich, M.C. McCann, R. Ming, D.G. Peterson, M. Rahman, D. Ware, P. Westhoff, K.F.X. Mayer, J. Messing and D.S. Rokhsar. 2009. The *Sorghum bicolor* genome and the diversification of grasses. Nature 457: 551–556.

Piperidis, G., M.J. Christopher, B.J. Carroll, N. Berding and A. D'Hont. 2000. Molecular contribution to selection of intergeneric hybrids between sugarcane and the wild species *Erianthus arundinaceus*. Genome 43: 1033–1037.

Piperidis, N., J.-w. Chen, H.-h. Deng, L.-P. Wang, P. Jackson and G. Piperidis. 2010. GISH characterization of *Erianthus arundinaceus* chromosomes in three generations of sugarcane intergeneric hybrids. Genome 53: 331–336.

Price, H.J., G.L. Hodnett, B.L. Burson, S.L. Dillon and W.L. Rooney. 2005. A *Sorghum bicolor* x *S. macrospermum* hybrid recorvered by embryo rescue and culture. Aust. J. Bot. 53: 579–582.

Price, S. 1965. Interspecific hybridization in sugarcane breeding. Proc. Int. Soc. Sugar Cane Technol. 12: 1021–1026.

Ram, B., T.V. Sreenivasan, B.K. Sahi and N. Singh. 2001. Introgression of low temperature tolerance and red rot resistance from *Erianthus* in sugarcane. Euphytica 122: 145–153.

Ram, S.G., S.H. Ramakrishnan, V. Thiruvengadam and J.R.K. Bapu. 2008. Prefertilization barriers to interspecific hybridization involving *Gossypium hirsutum* and four diploid wild species. Plant Breeding 127: 295–300.

Ramsey, J. and D.W. Schemske. 1998. Pathways, mechanisms, and rates of polyploid formation in flowering plants. Annu. Rev. Ecol. Syst. 29: 467–501.

Roach, B.T. 1972. Nobilisation of sugarcane. Proc. Int. Soc. Sugar Cane Technol. 14: 206–216.

Rodrangboon, P., P. Pongtongkam, S. Suputtitada and T. Adchi. 2002. Abnormal embryo development and efficient embryo rescue in interspecific hybrids, *Oryza sativa* x *O. minuta* and *O. sativa* x *O. officinalis*. Breeding Sci. 52: 123–129.

Sierra, R., A. Smith, C. Granda and M.T. Holtzapple. 2008. Producing fuels and chemicals from lignocellulosic biomass. Chem. Eng. Prog. 104: S10–S18.

Sobral, B.W.S., D.P.V. Braga, E.S. LaHood and P. Keim. 1994. Phylogenetic analysis of chloroplast restriction enzyme site mutations in the Saccharinae Griseb. subtribe of the Andropogoneae Dumort. tribe. Theor. Appl. Genet. 87: 843–853.

Somerville, C., H. Youngs, C. Taylor, S.C. Davis and S.P. Long. 2010. Feedstocks for lignocellulosic biofuels. Science 329: 790–792.

Sreenivasan, T.V., B.S. Ahloowalia and D.J. Heinz. 1987. Cytogenetics. pp. 211–253. *In*: D.J. Heinz (ed.). Sugarcane Improvement through Breeding, vol. 11. Elsevier, Amsterdam.

Stevenson, G.C. 1965. Genetics and breeding of sugar cane. Longman, London, UK.

Waclawovsky, A.J., P.M. Sato, C.G. Lembke, P.H. Moore and G.M. Souza. 2010. Sugarcane for bioenergy production: an assessment of yield and regulation of sucrose content. Plant Biotechnol. J. 8: 263–276.

Zhang, Y.H. 2008. Reviving the carbohydrate economy via multiproduct lignocellulose biorefineries. J. Ind. Microbiol. Biot. 35: 367–375.

The Cell Wall Architecture of Sugarcane and its Implications to Cell Wall Recalcitrance

Marcos S. Buckeridge,[1,]* *Wanderley D. dos Santos,*[2]
Marco A.S. Tiné[3] *and Amanda P. de Souza*[1]

ABSTRACT

Developing processes to produce bioethanol from sugarcane
biomass residues such as bagasse and leaves can significantly
increase the production of bioethanol. This process is named
2nd generation (2G) as opposed to the 1st generation (1G)
process (based on sucrose). The latter is well established in
Brazil nowadays, with ca. 400 mills currently in operation in
this country. The problem of recalcitrance of the cell wall to
hydrolysis has prevented rapid development of 2G processes due
to the lack of knowledge about sugarcane cell wall complexity.
In this chapter, we propose strategies to approach the problem
of recalcitrance by use of features related to cell wall complexity.

[1] Laboratory of Plant Physiological Ecology (LAFIECO), Department of Botany, Institute of
Biosciences, University of São Paulo.
[2] Laboratory of Plant Biochemistry, University of Maringá, PR, Brazil.
[3] Núcleo de Fisiologia e Bioquímica de Plantas, Instituto de Botânica de São Paulo.
* Corresponding author: msbuck@usp.br

The architecture of sugarcane cell walls is discussed using available information and some new data using microscopy. A cell wall architectural model is used to propose a strategy for hydrolysis capable of attacking cell walls more efficiently. The model proposes to overcome the three major limiting steps: the cell wall porosity, the glycomic code of hemicelluloses and the disassembly of the macrofibrillar structure. We also highlight the importance of cell wall architecture diversity in different tissues of sugarcane plants as a future target to lead 2G processes to the same level of efficiency as the 1G bioethanol processes that Brazil currently has. The knowledge status about sugarcane cell wall architecture will also help open the way for development of 2G technologies for other bioenergy grasses.

Introduction

During the 20th century, sugarcane became one of the major bioenergy sources in the planet, being the biomass used for production of almost half of the bioethanol. The reason for this is that bioethanol production from the sucrose accumulated in stems of sugarcane (1st generation bioethanol—1G) became highly efficient and cost effective in the southeast of Brazil during the last third of the 20th century (De Souza et al. 2014), consolidating the establishment of more than 400 mills in the country (Amorim et al. 2011). This happened due to the actions taken by Brazil in the 1970s to cope with the oil crisis, which led to measures that culminated in a large increase in sugarcane productivity. At the same time, there was a dramatic increase in the efficiency of sugar extraction, fermentation and distillation processes during a period of approximately 20 years, from 1980 to 1998 (Goldemberg 2007; Goldemberg 2010).

Even after becoming so efficient in many ways in the production of 1G bioethanol, it is possible to increase it even more. According to some authors, the increase in sucrose content or the enhancement of biomass in Brazilian varieties of sugarcane is achievable, since this crop plant has yet to reach the theoretical maximum of its productivity (Hotta et al. 2010; Waclawovsky et al. 2010). Furthermore, the biomass used for bioethanol production could be increased significantly if non-sucrose containing plant materials—cell walls, present in bagasse and leaves of sugarcane—could be used for bioethanol production through the process called 2nd generation (2G) (Buckeridge et al. 2010; Dos Santos et al. 2011; Buckeridge et al. 2012). Second generation bioethanol production from sugarcane in Brazil is important because it can, potentially, increase up to 40% of the bioethanol output in the country, without increasing the planted area (Buckeridge et al. 2010).

The 2G bioethanol technology has been a central focus of bioenergy researchers in recent years. The biochemical route, i.e., the use of enzymes from microorganisms to hydrolyze cell walls has been the main route of choice because of the mild conditions in which it occurs and the potential of control it provides. In this chapter, we examine some points related to cell wall architecture that we consider fundamental to the advance in cell wall hydrolysis: the biochemical process of polysaccharide digestion. This step imposes one of the most important barriers to reach economically viable 2G bioethanol production from sugarcane in Brazil.

Hydrolysis: A Key Step for 2G Bioethanol Production

In order to gain access to the sugars present in cell walls, one needs to be able to break polysaccharides into their basic units—the monosaccharides. This can be accomplished by a process called hydrolysis. However, plant cell wall components are resistant or recalcitrant to hydrolysis as a result of 500 million years of coevolution among plants, pathogens and herbivores. The reason why hydrolysis of polymers to monosaccharides is a key for 2G bioethanol technology is because bioethanol production is made from fermentation of these simple sugars (Amorim et al. 2011).

The problem of resistance to hydrolysis has been scientifically addressed mainly by using different fungal and bacterial enzymes capable of attacking different polysaccharides present in plant cell walls (Pauly and Keegstra 2008; dos Santos et al. 2011). Approaches vary from ones that focus on finding and characterizing individual enzymes (see review by Yang et al. 2011) to those that use enzyme cocktails to degrade biomass (e.g., Walton et al. 2011; Mohanram et al. 2013).

Cocktails of enzymes are usually added to biomass of plants such as poplar, maize, miscanthus and sugarcane (Taherzadeh and Karimi 2007; Yang et al. 2011), and the release of sugars is subsequently measured. Conversely, several publications investigated the causes for lignocellulose recalcitrance from a point of view of plant cell wall complexity. Most publications seldom consider the polysaccharide fine structure or polymer assembly as an architectural entity that can interfere with the attack of enzymes to biomass.

Despite the fact that maize, one of the most important biomasses for use as a bioenergy source, had its cell walls deeply studied (Carpita 1984), relatively little is known about grasses like miscanthus, switchgrass, sorghum and sugarcane. Only recently, cell wall structure and architecture of sugarcane and miscanthus have been more deeply studied and associated with cell wall degradation for bioethanol production (De Souza et al. 2013;

De Souza et al. 2015). This lack of knowledge about cell walls contrasts drastically with the relatively abundant information being produced about enzymes. This is likely to be one of the main factors that prevent the formulation of more balanced and precise hypotheses about the causes for synergies and complex relationships among enzymes.

Sugarcane Wall Complexity at Different Scales

Cell wall complexity can be approached didactically by distinguishing three levels of organization: (1) monosaccharide composition, (2) architectural level and (3) architectural diversity (Fig. 1).

The first and mostly used organization level in bioenergy research is the monosaccharide composition of the cell wall (Fig. 1A). The analysis of monosaccharide composition provides information that can indeed help to partially interpret the pattern of action of enzymes either by examining the composition of the biomass left intact or by measuring the amounts and types of monosaccharides released after hydrolysis. However, in terms of polysaccharide composition, there is a limitation if one uses only monosaccharide measurements. This is because it is very difficult to use monosaccharide quantification to infer the structure of complex polysaccharides, even when linkage information about them is available. In grasses, for instance, glucose is found in cellulose, but it is also found in mixed linkage β-glucan and xyloglucan; arabinose is found in arabinoxylan, but is also present in pectins.

Another relevant limitation is related to the fact that saccharification has been usually gauged by the release of glucose from biomass treated with isolated enzymes or cocktails. This approach evaluates the result of enzyme action only on part of the wall. In other words, by using the release of glucose after pretreatments, researchers mostly measure the attack of enzymes to cellulose, since cellulases are the preferred enzymes used (Gomez et al. 2010; Walton et al. 2011), leaving aside the analysis of hemicelluloses and pectins that, in sugarcane, are quantitatively relevant.

The second level of complexity is related to the assembly of polymers in the wall (Fig. 1A). This is called cell wall architecture. McCann and Roberts used the term in 1991, when they proposed a model of independent domains of the wall. Later on, in 1993, Carpita and Gibeaut proposed the existence of two types of wall that they named Type I (xyloglucan-rich) and Type II (arabinoxylan-rich), which has been subsequently extended with the discovery of Type III (mannan-rich) in Pteridophytae (Silva et al. 2011). Sugarcane cell walls consist of Type II walls, as recently determined by De Souza et al. (2013).

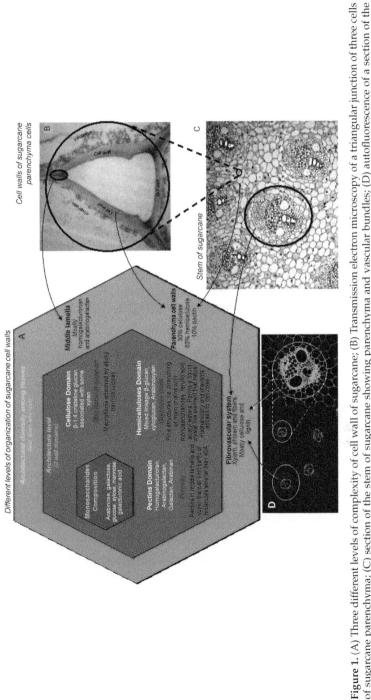

Figure 1. (A) Three different levels of complexity of cell wall of sugarcane; (B) Transmission electron microscopy of a triangular junction of three cells of sugarcane parenchyma; (C) section of the stem of sugarcane showing parenchyma and vascular bundles; (D) autofluorescence of a section of the stem of sugarcane showing vascular bundles. Right = blow up showing a vascular bundle with increased concentration of lignin in xylem and fibers (Photos by Debora C.C. Leite and Sávio S. Ferreira).

The third level of complexity is related to the diversity of combinations of wall domains in different tissues of a plant (Fig. 1A–D). Although the average composition of cell wall in a single species is often quite uniform, there is a great diversity among cell types (Pauly and Keegstra 2010). This diversity can be related to different ratios or qualitative differences in wall components. Regarding composition, sugarcane walls have been shown as relatively conserved among organs. Stems, leaves and roots are composed of cellulose, hemicelluloses (arabinoxylan, mixed-linkage β-glucan, xyloglucan and traces of mannan) and pectins (homogalacturonan, rhamnogalacturonan I branched with short chains of 1,3-1,6-β-galactan and 1,5-α-arabinan) (De Souza et al. 2013; Leite 2012). However, the three main classes of wall carbohydrate polymers (cellulose, hemicelluloses and pectins) and lignin are arranged in the wall in slightly different ways, changing in accordance with the tissue and the cell type (Fig. 1A–D).

In this chapter we argue that cell wall architecture is key to the production of 2G bioethanol. Also, a more detailed analysis of the three cell wall polysaccharide domains of sugarcane is given. In this way we intend to show how polysaccharide composition and their arrangement in the sugarcane walls can interfere with the hydrolysis process.

The Main Cell Wall Domains in Sugarcane and their Implications to Hydrolysis

Cellulose Domain

Cellulose *micro*fibrils, formed by 36 glucan chains, are the crystalline element of the cell wall composite and are essential for the mechanical properties of the cells and tissues in plants. In the fibrovascular system, for example, cellulose along with some hemicelluloses (xylan and/or xyloglucan) strongly bound to its surface, are usually associated with the higher lignification levels of the tissue, increasing mechanical resistance of fibers and vascular bundles.

For maize cell walls, Ding and Himmel (2006) produced microscopic evidence that supports a model in which seven *micro*fibrils are arranged in the wall forming *macro*fibrils. After examining sugarcane walls by the use of direct visualization with Atomic Force Microscopy (AFM), we found very similar *macro*fibrils to the ones found in maize by Ding and Himmel (2006). These were observed in native walls from parenchyma cells (Figs. 2A and 2B) as well as after treatment with xylanase (Fig. 2C), when

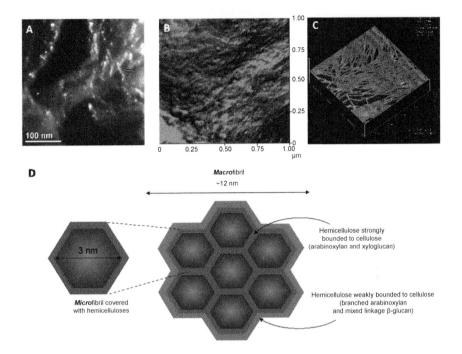

Figure 2. Cell walls of sugarcane under (A) Electron-diffraction microscopy; (B) Atomic Force microscopy (AFM); (C) AFM showing parenchyma cell walls treated with endo-β-xylanase, exposing the *macro*fibrils. *Macro*fibrils (12nm diameter) are pointed by red arrows. Walls display amorphous materials that possibly correspond to hemicelluloses and pectins; (D) hypothetical model for the arrangement of *macro*fibrils in sugarcane according to the proposition of Ding and Himmel (2006) for maize walls (Photos acknowledge participation of Vinicius L. Pimentel, Jefferson Bettini and Rodrigo V. Portugal).

naked *macro*fibrils could be clearly visualized. In both cases, the average diameter of the *macro*fibrils was ca. 12 nm, which is consistent with the assembly of a bundle containing seven *micro*fibrils with approximately 3 nm each (Fig. 2D).

Here we extend Ding and Himmel's model by including possible hemicelluloses that might be associated with this domain (Fig. 2D). Although there is some evidence that mixed linkage β-glucans could bind directly to cellulose in cell walls of grasses (Kiemle et al. 2014), it is likely that in sugarcane, xyloglucan and arabinoxylan would be the main polymers tying together the 3 nm *micro*fibrils. This proposal is made on the basis of the evidence that xyloglucan and arabinoxylan are more strongly bound to cellulose than mixed linkage β-glucans in sugarcane walls—evidence obtained by the fractionation with alkali followed by treatment with restriction endo-hydrolases as well as with monoclonal antibodies (De Souza et al. 2013). It is also noticeable that arabinoxylans are attached to the less soluble fraction of the wall from sugarcane. Thus, it can be speculated that arabinoxylans might be associated with the surface of both the *macro-* and *micro*fibrillar complex (Fig. 2D).

If such a model system could be confirmed by more experimental evidence, the cellulose-xyloglucan-arabinoxylan *macro*fibrillar complex might be considered as a sort of "basic unit" for the wall of sugarcane cellulose domain.

As shown by Electron Diffraction Microscopy and the AFM pictures in Figs. 2A and 2B respectively, this "basic unit" is apparently embedded in an independent amorphous matrix. In other words, the "basic unit" would be surrounded by a coating layer of more branched xyloglucan and arabinoxylan that are embedded into a more soluble matrix of pectins and mixed linkage β-glucan. Evidence in favor of this hypothesis is that these polymers are indeed more soluble and easily extracted during cell wall fractionation steps as described by De Souza et al. (2013). A schematic representation of this model can be seen in Fig. 3B in which several bundles of *macro*fibrils are immersed in a matrix of more soluble polymers. Furthermore, the cellulose domain in sugarcane seems to be strongly associated with lignification. As can be seen in Fig. 1D, strong autofluorescence appears in the fibrovascular bundles whereas lighter autofluorescence can be seen in the parenchyma. The latter might be

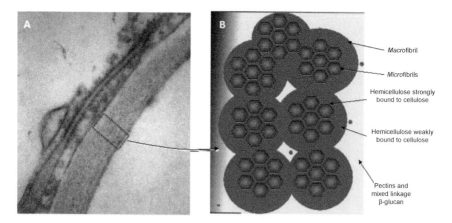

Figure 3. (A) Transmission electron micrograph showing the wall of a cell from the parenchyma of sugarcane; (B) Proposed model for the average architecture of sugarcane cell wall. The representation is shown as a transversal section of the wall.

related to the occurrence of ferulic acid branchings of arabinoxylans and xyloglucans (see below).

Driemeier et al. (2012) confirmed that the crystallites of cellulose in sugarcane stem are broadly consistent with a 24 to 36-chains of elementary fibrils of cellulose. These authors also observed that the average *microfibrillar* angles varied considerably with the position in the tissue as well as with developmental stage of sugarcane. Assuming that most of the walls in sugarcane occur in the form of *macrofibrils*, it is possible to speculate that the area available to attack by enzymes (e.g., cellulases) is relatively small in comparison to what it would be if only *microfibrils* were present. This may be quite significant for hydrolysis of sugarcane materials, since it has been reported by Bragatto et al. (2012) that surface availability and porosity of different cellulosic materials correlate positively with endo-glucanase attack.

Hemicelluloses Domain

Hemicelluloses form a domain of complex and entangled polysaccharides that are more hydrated than cellulose and, from a mechanical point of view, behave like the molecules of the pectin domain (see below). The hemicellulose domain surrounds cellulose fibrils and its polymers display a large variety of branches.

Sugarcane hemicelluloses are composed of arabinoxylan (ca. 60%), xyloglucan (ca. 10%), mixed linkage β-glucan (ca. 10%) and mannan (trace amounts) (De Souza et al. 2013). They are very complex branched polysaccharides that are thought to be self-associated to a certain extent as well as associated with the surface of cellulose. On the basis of the highly complex branching patterns of hemicellulose with sugars, acetyl or methyl esters and ferulic acid, the existence of a glycomic code has been recently proposed to exist in plant cell walls (Buckeridge and De Souza 2014).

The glycomic code in hemicelluloses of sugarcane walls has been assessed with the use of restriction hydrolases. The restriction activity (in the sense that the hydrolase acts as a restriction enzyme on DNA molecules) appears when many—if not all—of the endo-hydrolases such as cellulases, xyloglucanases, lichenases and mannanases are used in low concentrations to produce oligosaccharides that are diagnostic of the fine structure in polysaccharides (Buckeridge and De Souza 2014).

Two important features of the glycomic code in hemicelluloses are thought to be associated with resistance to hydrolysis in sugarcane: (1) the polymer-polymer associations and (2) resistance to hydrolysis due to the position of the branching motifs in arabinoxylan and xyloglucan that impede full access of the enzyme to the main chain. Evidence in favor of the hypothesis that polymer-polymer associations are part of the mechanism of resistance to hydrolysis is the fact that when intact cell walls (i.e., alcohol insoluble residue samples of sugarcane) are treated with hemicellulose-specific endo-glycanases such as endo-β-xylanase and endo-β-xyloglucanase, only a small percentage of the possible oligosaccharides retrievable from these polymers is released during the reaction (Figs. 4A and 4C). This contrasts with what happens when xyloglucans and arabinoxylans are extracted from sugarcane walls by fractionation with alkali (Figs. 4B and 4D, De Souza et al. 2013). On the other hand, this may not be the case for some of the mixed-linkage β-glucan in the wall, especially the one present in the culms of sugarcane, which is mostly released from intact walls (Figs. 4E and 4F, see De Souza et al. 2013 for more details).

Thus, at the hemicellulose domain level, the resistance to hydrolysis in sugarcane appears to be given by a combination of polymer-polymer interactions along with the patterns of branching of the main chains of polysaccharides, except for the case of the mixed linkage β-glucan. One particular feature that seems to be quite important as a barrier to main chain hydrolysis is the presence of decorations with acetyl esters and ferulic acid. Also, as the latter is a compound known to initiate lignin polymerization, it is reasonable to think that lignin can cross-react with polymers in the wall of sugarcane, turning them almost completely resistant to hydrolysis. This situation seems to be characteristic for cells of the vascular bundles (xylem and fibers) and relatively infrequent in tissues like the parenchyma of the stem (Fig. 1D).

Pectins Domain

The pectin domain is formed by polymers that are more soluble in water. It is known to be associated with wall porosity and it is chemically independent of the other domains. It is also possible that pectins coexist in the intercellular spaces together with hemicelluloses (McCann and Roberts 1991). The pectin domain is thought to be related to defense mechanisms in plants, being the first barrier that microorganisms have to face during cell invasion.

Pectins in sugarcane occur both in the cell walls and the middle lamella. Typical pectic polymers have been detected in all organs of sugarcane: homogalacturonan, arabinogalactan I, which may be branched with chains 1,3-1,6-β-galactan and 1,5-α-arabinan, the latter two polymers being probably short chains that branch with rhamnoglacturonan I (RGI) (De Souza et al. 2013; Leite 2012).

Immunolocalization of homogalacturonan in sugarcane root tissues shows that this polymer is present in the middle lamella (Fig. 5) whereas

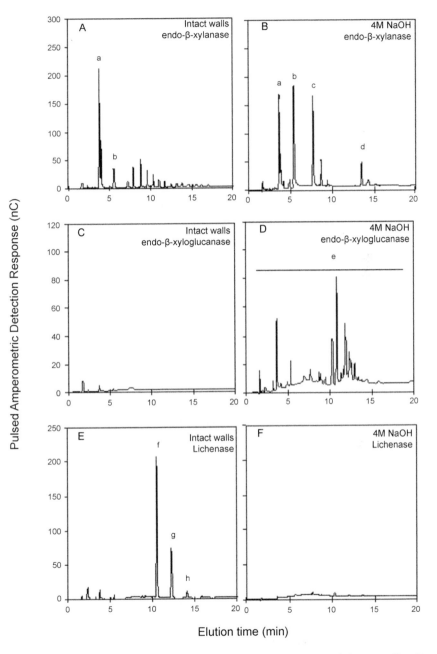

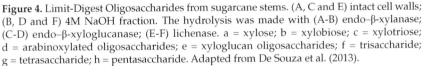

Figure 4. Limit-Digest Oligosaccharides from sugarcane stems. (A, C and E) intact cell walls; (B, D and F) 4M NaOH fraction. The hydrolysis was made with (A-B) endo–β-xylanase; (C-D) endo–β-xyloglucanase; (E-F) lichenase. a = xylose; b = xylobiose; c = xylotriose; d = arabinoxylated oligosaccharides; e = xyloglucan oligosaccharides; f = trisaccharide; g = tetrasaccharide; h = pentasaccharide. Adapted from De Souza et al. (2013).

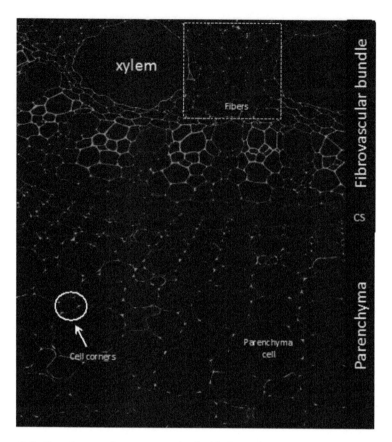

Figure 5. Section of a root of sugarcane stained with the antibody JIM7. This antibody stains strongly methyl esterified homogalacturonan (MS-HG) with some background staining of rhamnogalacturonan I (RGI). MS-HG is more concentrated in the cell corners of parenchyma cells and in the cells surrounding the xylem. Staining is also in the cell wall and middle lamella in this tissue. Note that in fibers, staining is seen only in the cell corners but not in the walls that in this case display architecture with cellulose domain mainly. The caspary stripes (CS), which have lignified walls, do not show staining with the antibody (Photo by Debora C.C. Leite and Utku Avci).

RGI, galactan and arabinan seem to be found in the walls of all cells (Leite 2012). Cell corners are the richest regions of cell walls of parenchyma containing homogalacturonan whereas other tissues, like those in the fibrovascular bundles, display some diversity of distribution of pectins (Fig. 5, Leite 2012). Whereas the cells surrounding xylem appear to have relatively more pectin in cell walls, in fibers only the middle lamella and especially cell corners contain pectin (Fig. 5). The walls in this case are thick and probably richer in elements of the cellulose domain. Here we speculate that stem and leaf tissues, that display very similar fundamental structures (fibrovascular bundles immersed in a parenchyma tissue), would probably have similar pectin distribution throughout their walls.

The main interference of the pectin domain in hydrolysis appears to be the fact that pectins play a major role in the determination of cell wall porosity, a feature that correlates directly with the extent to which enzymes can penetrate biomass. In order to hydrolyze the supramolecular complex of polymers of the cell walls, enzymes should be able to access the composite by attacking its surface and/or penetrating it.

The porosity of plant cell walls have been measured in a few cases (Carpita 1982; Baron-Epel et al. 1988), suggesting that the average pore size of the wall matrix limits the traffic of molecules with diameters larger than 35–40Å (Table 1). In sugarcane, cell wall porosity has been studied by thermoporometry leading to pores sizes that vary among different sugarcane tissues (parenchyma, bundles and rind) at different positions in sugarcane culm (top, medium and bottom heights) (Maziero et al. 2013). These authors concluded that the porosity of sugarcane biomass particles

Table 1. Examples of cell wall hydrolases associated with the three levels of cell wall architecture that had their structure characterized. Many proteins are near or within the average pore-size range (50–200Å) of the wall of sugarcane, except for β-expansin*.

Enzyme	Organism	Lattice parameters (Angstroms)			Reference
		a	b	c	
Wall Porosity					
Endopolygalacturonase	*Aspergillus niger*	65.5	201.24	49.07	Van Santen et al. (1999)
Pectin methyl esterase	Carrot	49.5	77.6	89.2	Johansson et al. (2002)
Glycomic Code					
α-galactosidase	Rice	63.7	71.4	84.2	Fujimoto et al. (2003)
β-galactosidase	*Trichodermareseei*	67.4	69.2	81.5	Maksimainen et al. (2011)
XTH**	Nasturtium	116.1	116.1	63.1	Bauman et al. (2007)
Lichenase	Barley	49.6	82.9	77.5	Muller et al. (1998)
β-expansin*	Maize	**35**	**30**	**24**	Yennawar et al. (2006)
Cellulose					
β-glucosidase	Maize	60	118	70	Czjzek et al. (2001)
Cellobiohydrolase I	*Trichoderm areseei*	60	50	40	Divine et al. (1994)

**Xyloglucan endo-β-transglycosilase

varied from 1 to 400 nm, with averages ranging between 5 and 20 nm (i.e., 50 to 200Å) depending of the tissue and position analyzed.

Since most of the hydrolases that have resolved structures display diameters above 50Å, it can be speculated that penetration into the intact cell wall matrix by hydrolases might be partially limited. Nevertheless, this hypothesis is still to be tested directly on sugarcane biomass.

Solving Challenges of Cell Wall Recalcitrance Related to Polysaccharides

Considering what is known about sugarcane cell wall composition and the architectural model presented in this chapter, the proposition of enzyme combinations that would be capable of hydrolyzing plant cell walls

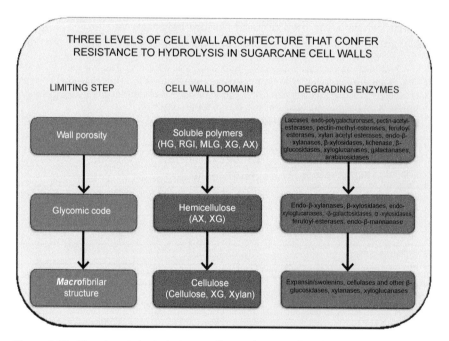

Figure 6. Limiting steps to hydrolysis according to the type of recalcitrance (blue) imposed by different domains of the cell wall (red) of sugarcane. The enzymes associated with every respective step are listed on the right (green). HG = homogalacturonan; RGI = rhamnogalacturonan I; MLG = mixed linkage-β-glucan; XG = xyloglucan; AX = arabinoxylan.

completely will have to take into account three major limiting steps: (1) wall porosity; (2) glycomic code and (3) *macro*fibrillar structure (Fig. 6). In each case, a different combination of wall degrading enzymes would have to be used in spite of the fact that some of the classes of polysaccharides are similar.

Independently of polysaccharides hydrolysis, lignin should be attacked. This can be achieved with several types of pretreatments (Soccol et al. 2010) as well as with biological strategies that decrease lignin in sugarcane cell walls (Jung et al. 2013). Although it forms a separate non-carbohydrate polymer, lignin is attached to polysaccharide branching residues (feruloylated arabinosyl residues in arabinoxylan for instance), which confers hydrophobicity to the wall, preventing hydrolysis.

Regarding cell wall polysaccharides, a first cocktail would have to be directed to increase wall porosity (Fig. 6). Therefore, this cocktail should have enzymes capable of digesting pectins and mixed linkage β-glucans. In non-pretreated sugarcane biomass materials, enzymes capable to attack lignin (e.g., laccases) would have to be added in order to open access to the hemicellulose's branching decorations (i.e., the glycomic code). This first attack will also confront some of the more soluble xyloglucans and arabinoxylans so that enzymes specific to highly branched forms of these polymers would have to be present.

Once porosity is increased by taking out soluble polymers, a second enzyme cocktail should be able to break polymer-polymer interactions as well as to degrade highly complex hemicelluloses, the most abundant one being arabinoxylan (Fig. 6). This polymer is thought to be self-interacting through diferulic bonds and in this regard, feruloyl-esterases are extremely relevant to break the glycomic code of the hemicellulose domain.

The third and last step will be the degradation of the *macro*fibrils. It should be noted here that we have hypothesized that strongly bound hemicelluloses belonging to the classes of xylans and xyloglucans are the adhering elements that maintain *micro*fibrils together in sugarcane. Thus, in order to have access to the surface of *macro*fibrils, swollenins/expansins will also have to act, peeling off the hemicellulose coating and opening the way (and greatly increasing surface) for the attack of cellulases to *micro*fibrils.

It must be highlighted that this is only a theoretical approach and that experiments are needed to either corroborate or not the assembly of these three-cocktails-strategy. Even so, theoretically, such a procedure might succeed to yield relatively more sugars than the enzyme cocktails currently proposed/used, since the latter hardly target important elements related to porosity (e.g., pectins and mixed linkage β-glucans) and may also lack some key elements related to the glycomic code.

Architectural Diversity Among Tissues: a Step for the Future

Besides the complexity of cell wall architecture, there is a higher level of organization that will have to be addressed in order to obtain even higher efficiency of hydrolysis: the diversity of combinations among cell wall architectural elements in different cells of different tissues. The complexity related to all these variations and the importance of that for the bioenergy sector have been previously addressed by Pauly and Keegstra (2010).

The diversity aspect in 2G bioethanol technology is important because it is known that sugarcane stems, for example, have different wall compositions depending on the height of the stem (Costa 2012) or the type of cell (Driemeier et al. 2012). Also, the leaves left in the field have been cogitated as one of the possible sources of biomass for bioethanol production (Hari Krishna et al. 1998; Krishnan et al. 2010). Leaves have proportionally more vascular bundles in relation to stems (De Souza et al. 2013), which makes them relatively rich in the cellulose domain. This may require proportionally less addition of the enzymes related to the porosity level of recalcitrance, but a reinforcement of the enzymes that attack the cellulose domain. In this case, the level of redundancy of walls in different tissues of sugarcane becomes relevant to develop optimal hydrolysis processes (Buckeridge 2006).

In scenarios like the ones proposed by Macrelli et al. (2012), in which bagasse and leaves are integrated as a 2G process coupled to the 1G, the knowledge of architectural diversity for the cell walls of sugarcane in different tissues would probably be significant to design procedures capable to extract most of the sugars present in this plant.

Although the features stated above are related to very specific details of the walls, they are very likely to help improving 2G technology since they should pave the way to increase the efficiency of cell wall hydrolysis. Furthermore, it must be highlighted that in addition to the benefits that knowledge on sugarcane cell wall architecture can bring to development of 2G in Brazil, it could also be applied to other bioenergy grasses, like miscanthus, sorghum, maize and switchgrass.

Acknowledgements

The authors acknowledge the financial support of the Instituto Nacional de Ciência e Tecnologia do Bioetanol – INCT do Bioetanol (FAPESP 2008/57908-6 and CNPq 574002/2008-1) and of the Centro de Processos Biológicos e Industriais para Biocombustíveis—CeProBIO (FAPESP 2009/52840-7 and CNPq 490022/2009-0). The authors also like to thank Vinícius L. Pimentel, Jefferson Bettini and Rodrigo V. Portugal (LNNano-CNPEM, Campinas) for the help with the preparation of materials for AFM analyses, and Débora C.C. Leite and Savio S. Ferreira for the photos from microscopy. Pectin staining with JIM 7 was possible thanks to the collaboration with Michael Hahn and Utku Avci from the CCRC-Georgia.

References

Amorim, H.V., M.L. Lopes, J.V.C.O. Oliveira, M.S. Buckeridge and G.H. Goldman. 2011. Scientific challenges of bioethanol production in Brazil. Applied Microbiology and Biotechnology 91: 1267–1275.

Baron-Epel, O., P.K. Gharyal and M. Shindler. 1988. Pectins as mediators of wall porosity in soybean cells. Planta 175: 389–395.

Baumann, M.J., J.M. Eklof, G. Michel, A.M. Kallas, T.T. Teeri, M. Czjzek and H. Brumer, III. 2007. Structural evidence for the evolution of xyloglucanase activity from xyloglucan endo-transglycosylases: biological implications for cell wall metabolism. Plant Cell 19: 1947–1963.

Bragatto, J., F. Segato, J. Cota, D.B. Mello, M.M. Oliveira, M.S. Buckeridge, F.M. Squina and C. Driemeier. 2012. Insights on how the activity of an endoglucanase is affected by physical properties of insoluble celluloses. The Journal of Physical Chemistry B 116: 6128–6136.

Buckeridge, M.S. 2006. Implications of emergence degeneracy and redundancy for the modeling of the plant cell wall. pp. 41–47. In: T. Hayashi (Org.). The Science and the Lore of the Plant Cell Wall: Biosynthesis Structure and Function. Boca Raton: Brown Walker Press.

Buckeridge, M.S. and A.P. De Souza. 2014. Breaking the "Glycomic Code" of cell wall polysaccharides may improve second-generation bioenergy production from biomass. Bioenergy Research. DOI 101007/s12155-014-9460-6.

Buckeridge, M.S., W.D. Santos and A.P. De Souza. 2010. Routes for cellulosic ethanol in Brazil. pp. 365–380. In: L.A.B. Cortez (Org.). Sugarcane bioethanol: R&D for productivity and sustainability. Edgard Blucher, São Paulo.

Buckeridge, M.S., A.P. Souza, R.A. Arundale, K.J. Anderson-Teixeira and E. DeLucia. 2012. Ethanol from sugarcane in Brazil: a "midway" strategy for increasing ethanol production while maximizing environmental benefits. Global Change Biology and Bioenergy 4: 119–126.

Carpita, N.C. and D.M. Gibeaut. 1993. Structural models of primary cell walls in flowering plants: consistency of molecular structure with the physical properties of the cell wall during growth. Plant Journal 3: 1–30.

Carpita, N.C. 1984. Fractionation of hemicelluloses from maize cell walls with increasing concentrations of alkali. Phytochemistry 23: 1089–1093.

Carpita, N.C. 1982. Limiting diameters of pores and the surface structure of plant cell walls. Science 218: 813–814.

Costa, T.H.F. 2012. Evaluation of recalcitrance at different internode regions derived from hybrid sugarcane with varying amounts of lignin. Master's dissertation from Engineering School of Lorena.

Czjzek, M., M. Cicek, V. Zamboni, W.P. Burmeister, D.V. Bevans, B. Henrissat and A. Esen. 2001. Crystal structure of a monocoltyledon (maize ZMGlu1) β-glucosidase and a model of its complex with *p*-nitrophenyl β-D-thioglucoside. Biochemical Journal 354: 37–46.

De Souza, A.P., D.C.C. Leite, S. Pattathil, M.G. Hahn and M.S. Buckeridge. 2013. Composition and structure of sugarcane cell wall polysaccharides: implications for second-generation bioethanol production. Bioenergy Research 6: 564–579.

De Souza, A.P., A. Grandis, D.C.C. Leite and M.S. Buckeridge. 2014. Sugarcane as a bioenergy source: history performance and perspectives for second-generation bioethanol. Bioenergy Research 7: 24–35.

De Souza, A.P., C.L.A. Kamei, A.F. Torres, S. Pattathil, J.G. Hahn, L.M. Trindade and M.S. Buckeridge. 2015. How cell wall complexity influences saccharification efficiency in Miscanthus sinensis. Journal of Experimental Botany DOI 10.193/jxb/erv183.

Ding, S.Y. and M.E. Himmel. 2006. The maize primary cell wall microfibril: a new model derived from direct visualization. Journal of Agriculture and Food Chemistry 54: 597–606.

Divine, M., J. Stahlberg, T. Reinikanen, L. Ruohonen, G. Petterson, J.K. Knowles, T.T. Teeri and T.A. Jones. 1994. The three dimensional crystal structure of the catalytic core of cellobiohydrolaseI from *Trichoderma reesei*. Science 265: 524–528.

Dos Santos, W.D., E.O. Gómez and M.S. Buckeridge. 2011. Bioenergy and the sustainable revolution. pp. 15–26. *In*: M.S. Buckeridge and G.H. Goldman (Org.). Routes to cellulosic ethanol. Springer: New York.

Driemeier, C., W.D. dos Santos and M.S. Buckeridge. 2012. Cellulose crystals in fibrovascular bundles of sugarcane culms: orientation size distortion and variability. Cellulose 19: 1507–1515.

Fujimoto, Z., S. Kaneko, M. Momma, H. Kobayashi and H. Mizuno. 2003. Crystal structure of rice alpha-galactosidase complexed with D-galactose. Journal of Biological Chemistry 278: 20313–20318.

Goldemberg, J. 2007. Ethanol for a sustainable energy future. Science 315: 808–810.

Goldemberg, J. 2010. The role of biomass in the world's energy system. pp. 3–14. *In*: M.S. Buckeridge and G.H. Goldman (Org.). Routes to Cellulosic Ethanol. Springer: New York.

Gomez, L.D., C. Witehead, A. Barakae, C. Halpin and S.J. McQueen-Mason. 2010. Automated saccharification assay for determination of digestibility in plant materials. Biotechnology for Biofuels 3: 23–35.

Hari Krishna, S., K. Pasanthi, G.V. Chowdary and C. Ayyanna. 1998. Simultaneous saccharification and fermentation of pretreated sugarcane leaves to ethanol. Process Biochemistry 33: 825–830.

Hotta, C.T., C.G. Lembke, D.S. Domingues, E.A. Ochoa, G.M.Q. Cruz, D.M. Melotto-Passarin, T.G. Marconi, M.O. Santos, M. Mollinari, G.R.A. Margarido, A.C. Crivellari, W.D. Santos, A.P. De Souza, A.A. Hoshino, H. Carrer, A. Souza, A.F. Garcia, M.S. Buckeridge, M. Menossi, M.A. Van Sluys and G.M. Souza. 2010. The biotechnology roadmap for sugarcane improvement. Tropical Plant Biology 3: 75–87.

Johansson, K., M. El Ahmad, R. Friemann, H. Jornvall, O. Markovic and H. Eklund. 2002. Crystal structure of plant pectin methylesterase. FEBS Letter 514: 243–249.

Jung, J.H., W. Vermerris, M. Gallo, J.R. Fedenko, J.E. Erickson and F. Altpeter. 2013. RNA interference suppression of lignin biosynthesis increases fermentable sugar yields for biofuel production from field-grown sugarcane. Plant Biotechnological Journal 11: 709–716.

Kiemle, S.N., X. Zhang, A.R. Esker, G. Toriz, P. Gatenholm and D.J. Cosgrove. 2014. Role of (1,3)(1,4)-β-glucan in cell walls: interaction with cellulose. Biomacromolecules 15: 1727–1736.

Krishnan, C., L.C. Sousa, M. Jin, L. Chang, B.E. Dale and V. Balan. 2010. Alkali-based AFEX pretreatment for the conversion of sugarcane bagasse and cane leaf residues to ethanol. Biotechnology and Bioengineering 107: 441–450.

Leite, D.C.C. 2012. Cell wall modifications during aerenchyma formation in sugarcane roots. Master's dissertation. Instituto de Biociências da Universidade de São Paulo, 106 p.

Macrelli, S., J. Mogensen and G. Zacchi. 2012. Techno-economic evaluation of a 2nd generation bioethanol production from sugar cane bagasse and leaves integrated with the sugar-based ethanol process. Biotechnology for Biofuels 5: 22–40.

Maksimainen, M., N. Hakulinen, J.M. Kallio, T. Timoharju, O. Turunem and J. Rouvinen. 2011. Crystal strucutres of Trichoderma ressei β-galactosidase reveal conformational changes in the active site. Journal of Structural Biology 174: 156–163.

Maziero, P., J. Jong, F.M. Mendes, A.R. Gonçalves, M. Eder and C. Driemeier. 2013. Tissue-specific cell wall hydration in sugarcane stalks. Journal of Agricultural and Food Chemistry 61: 5841–5874.

McCann, M.C. and K. Roberts. 1991. Architecture of the primary cell wall. pp. 109–129. *In*: C.W. Lloyd (ed.). The cytoskeletal basis of plant growth and form. Academic Press: New York.

Mohanram, S., D. Amat, J. Choudhary, A. Arora and L. Nain. 2013. Novel perspectives for evolving enzyme cocktails for lignocellulose hydrolysis in biorefineries. Sustainable Chemical Processes 1: 15.

Muller, J.J., K.K. Thomsen and U. Heinemann. 1998. Crystal structure of barley 1,3-1,4-beta-glucanase at 2.0-A resolution and comparison with *Bacillus* 1,3-1,4-beta-glucanase. Journal of Biological Chemistry 273: 3438–3446.

Pauly, M. and K. Keegstra. 2008. Tear down this wall. Current Opinion in Plant Biology 11: 233–235.

Pauly, M. and K. Keegstra. 2010. Plant cell wall polymers as precursors for biofuels. Current Opinion in Plant Biology 13: 305–312.

van Santen, Y., J.A. Benen, K.H. Schröter, K.H. Kalk, S. Armand, J. Visser and B.W. Dijkstra. 1999. 1.68-A crystal structure of endopolygalacturonase II from *Aspergillus niger* and identification of active site residues by site-directed mutagenesis. Journal of Biological Chemistry 274: 30474–30480.

Silva, G.B., M. Ionashiro, T.B. Carrara, A.C. Crivellari, M.A.S. Tiné, J. Prado, N.C. Carpita and M.S. Buckeridge. 2011. Cell wall polysaccharides from fern leaves: Evidence for a mannan-rich Type III cell wall in *Adiantum raddianum*. Phytochemistry 72: 2352–2360.

Soccol, C.R., L.P.S. Vandenberghe, A.B.P. Medeiros, S.G. Karp, M.S. Buckeridge, L.P. Ramos, A.P. Pitarelo, V. Ferreira-Leitão, L.M. Gottschalk, M.A. Ferrara, E.P. da Silva Bon, L.M. de Moraes, J.A. Araújo and F.A. Torres. 2010. Bioethanol from lignocelluloses: status and perspectives in Brazil. Bioresource Technology 101: 4820–4825.

Taherzadeh, M.J. and K. Karimi. 2007. Enzyme-based hydrolysis processes for ethanol from lignocellulosic materials: a review. Bioresources 2: 707–738.

Van Santen, Y., J.A. Benen, K.H. Schröter, S. Armand, J. Visser and B.W. Dijkstra. 1999. 1.68-A crystal strucuture of endopolygalacturonase II from Aspergillus niger and identification of active site residues by site-directed mutagenesis. The Journal of Biological Chemistry 274(43): 30474–30480.

Waclawovsky, A.J., P.M. Sato, C.G. Lembke, P.H. Moore and G.M. Souza. 2010. Sugarcane for bioenergy production: an assessment of yield and regulation of sucrose content. Plant Biotechnological Journal 8: 263–276.

Walton, J., G. Banerjee and S. Car. 2011. GENPLAT: an Automated platform for biomass enzyme discovery and cocktail optimization. Journal of Visualized Experiments 56: 3314.

Yang, B., Z. Dai, S.Y. Ding and C.E. Wyman. 2011. Enzymatic hydrolysis of cellulosic biomass Biofuels 2: 421–450.

Yennawar, N.H., L.-C. Li, D.M. Dudzinski, A. Tabuchi and D.J. Cosgrove. 2006. Crystal structure and activities of EXPB1 (Zea m1), a β-expansin and group-1 pollen allergen from maize. Proceedings of the National Academy of Science of the United States of America 40: 14664–14671.

Economic Issues for Sugarcane as a Biofuel Feedstock

*Luis A. Ribera** and *Henry Bryant*

ABSTRACT

Sugarcane is a tropical crop that is processed into raw sugar and molasses. In the U.S., sugarcane is planted and harvested in Hawaii, Florida, Louisiana, and Texas. The U.S. ethanol industry began to take shape in the late 1970s producing what was then called "gasohol" in response to a doubling of oil prices to nearly $30 per barrel. As a result of crude oil prices rising to nearly $40 per barrel in the early 1980s, the industry expanded rapidly and by the middle 1980s, there were an estimated 170 plants producing approximately 1.51 billion liters (400 million gallons) per year using mainly corn as feedstock. Currently, there are 204 ethanol plants in 29 states in the U.S. with a production capacity of 51 billion liters (13.5 billion gallons) per year and still using mainly corn as feedstock. There are three primary U.S. sugar policy instruments in place. First, a price support loan program makes nonrecourse loans to sugar processors

Texas AgriLife Extension Service, Texas A & M University, 2401 East Highway 83, Weslaco, TX 78596.
* Corresponding author

against sugar that is pledged as collateral. Second, a marketing allotment program limits the quantity of sugar that marketers are allowed to sell each year. Third, a tariff rate quota program sets stiff tariffs for sugar imports above specific thresholds for each individual trading partner. Growth of the U.S. ethanol industry is directly related to Federal and State policies and regulations. The main current policy is included in the Energy Independence and Security Act of 2007 that sets the amount of renewable fuel that should be blended to gasoline sold in the U.S. to 36 billion gallons by 2022, also known as RFS2. Among other provisions, the RFS2 sets mandatory blend levels for renewable fuels while also establishing greenhouse gas reduction criteria and methodology for calculating lifecycle GHG emissions. In this chapter, we will examine how this policy and other policy instruments affect the potential of sugarcane as feedstock for sugar or bioethanol. In the absence of a major overhaul of U.S. sugar policy, we conclude that sugar prices will likely remain only loosely associated with energy prices since sugarcane will be unlikely to be used extensively for biofuel production.

Introduction

Sugarcane is a tropical crop that is processed into raw sugar and molasses. In the U.S., sugarcane is planted and harvested in Hawaii, Florida, Louisiana, and Texas (Table 1). Sugarcane is a perennial crop that can be harvested 4 to 5 times before reseeding. Sugarcane production in the U.S. is reported on a fiscal year basis, as the harvest season in Florida, Hawaii and Texas generally runs from October through April. The harvest season in Louisiana,

Table 1. U.S. Sugarcane Area, Yield and Production 2009.

	Area harvested (ac)	Yield (tons/ac)	Production (1,000 tons)
Florida	390,000	36.1	14,079
Hawai	21,700	67.2	1,458
Louisiana	425,000	31.0	13,175
Texas	41,000	35.0	1,435
US	877,700	34.4	30,193

Source: USDA-NASS, Agricultural Statistics 2011; ac: acre.

the northern most growing U.S. area, generally runs from late September through late December or early January. In the 2009 season, the U.S. grew 877,700 acres (ac) of sugarcane (Table 1). U.S. sugarcane production has trended downward in recent years, down from 1.02 million acres in the 2001 season due mainly to weather problems, i.e., hurricanes in Louisiana and Florida.

Sugarcane Production Costs

The cost of sugarcane production is very similar among three of the four producing states. Shapouri et al. (2006) showed that the cost of production between Florida, Louisiana, and Texas ranges between 15.3 to 16.42 cents per pound of raw sugar, while the cost of production in Hawaii is around 18.37 cents per pound. The high cost of sugar production is one of the main reasons for the decline in sugarcane acres in Hawaii.

Cost of production budgets for plant cane, first year cane, ratoon cane, and second to fifth year cane, were constructed for Louisiana and Texas (Table 2). Input prices were taken from the 2012 Projected Louisiana Sugarcane Production Costs prepared by the LSU Ag Center and the 2012 Texas Crop Enterprise Budget prepared by the Texas AgriLife Extension Service at Texas A&M University. As mentioned above, both Louisiana and Texas sugarcane budgets look very similar with the main difference being

Table 2. Plant and Ratoon Sugarcane Cost of Production for Louisiana and Texas 2011.

	Plant Cane		Ratoon Cane	
	Texas	Louisiana	Texas	Louisiana
	$/acre			
Variable Costs				
Seed Cost	$148.00	$130.45		
Planting	$204.20	$373.49		
Pesticides and application	$179.12	$206.31	$187.89	$275.30
Tillage, fuel, repair, int. on OP. CAP. and labor	$91.35	$89.66	$52.12	$101.59
Irrigation	$152.50		$140.90	
Harvesting	$219.92	$265.13	$181.58	$265.13
Fixed Costs				
Machinery and equipment depreciation	$28.19	$47.41	$19.26	$53.57
Total	$1,023.28	$1,112.45	$581.75	$695.59

irrigation cost in Texas while Louisiana's cane is rain fed, and a higher cost for planting in Louisiana. Planting is usually done manually, hence the higher cost, in Louisiana while cheaper agricultural labor is readily available in South Texas where sugarcane is grown.

Assuming 35 tons/ac sugarcane yield and 200 lbs. of raw sugar per ton of cane for both Louisiana and Texas, the production cost for plant cane is 15.89 and 14.62 cents per pound of raw sugar, respectively. The production cost for ratoon cane is 9.94 and 8.31 cents per pound of raw sugar for Louisiana and Texas, respectively. Moreover, amortizing the seed and planting costs over the usual five-year production cycle for sugarcane showed an annual production cost of 11.13 cents per pound for Louisiana and 9.57 cents per pound for Texas. Processing costs including credits for molasses, bagasse and any other is around eight to nine cents per pound of raw sugar, which is comparable to the report by Shapouri et al. (2006).

Ethanol and Sugarcane

The U.S. ethanol industry began to take shape in the late 1970s producing what was then called "gasohol" in response to a doubling of oil prices to nearly $30 per barrel. As a result of crude oil prices rising to nearly $40 per barrel in the early 1980s, the industry expanded rapidly and by the middle 1980s, there were an estimated 170 plants producing approximately 1.51 billion liters (400 million gallons) per year using mainly corn as feedstock (Vander Griend 2006). However, by July 1986 the price of oil went back to $10 per barrel and the gasohol industry collapsed as costs were not competitive with gasoline at lower oil prices. Substantial industry consolidation followed, and the remaining firms began to focus on decreasing production costs. By the late 1990s, the costs of production (primarily due to larger plants realizing scale economies, reduced enzyme costs, and higher corn to ethanol conversion efficiency) for ethanol were competitive with gasoline. It should be noted that the blenders' tax credit, discussed later in this chapter, remained in place throughout the 1970s and 1980s, providing about the same amount of incentives until 2011 as was provided some thirty years ago.

Currently, there are 204 ethanol plants in 29 states in the U.S. with a production capacity of 13.5 billion gallons (RFA 2012). In addition, there are 10 plants either under construction or expanding. The U.S. is the largest ethanol producing country with 13.2 billion gallons in 2010, followed by Brazil, the European Union, China and India (Table 3). The primary feedstocks used to produce ethanol in the U.S. are grains (corn and grain sorghum), while it is sugarcane in most of the other countries such as Brazil, India and Central America. The process of making ethanol from grains has

Table 3. Top 5 Ethanol Producing Countries 2010.

	Millions of Gallons
United States	13,200
Brazil	7,228
European Union	1,170
China	578
India	379

Source: RFA, FAPRI

evolved such that the yield of ethanol per unit of grain (especially corn and grain sorghum) has risen while conversion costs have declined over the past decade.

Brazil has nearly perfected the process of converting sugarcane to ethanol over the past 30 years. Ethanol yields per acre are higher for sugarcane-based ethanol than any other currently available feedstock (Matsuoka et al. 2009; Eggleston 2010). Table 4 indicates the estimates of Brazilian and U.S. ethanol production from sugarcane by Ribera et al. (2007). The cost of production in Brazil is $1.22 per gallon of ethanol ($0.32 per liter) excluding capital cost (Chaves 2006). Chaves stated that the cost of production in 2005 was $0.89 per gallon ($0.24 per liter) with the exchange rate of three *Real* per U.S. dollar. However, due to the depreciation of the U.S. currency against the Brazilian *Real* to around 2.20 *Real* per U.S. dollar in 2006, the cost per gallon has increased to $1.22. The estimated total cost of production per gallon of ethanol from sugarcane in the U.S. is $1.87 ($0.49 per liter) (Ribera et al. 2007), assuming it costs $17/ton for cane.

Table 4. Estimated Costs of Production for Sugarcane-based Ethanol.

	Brazil[a]		U.S.	
	$/liter	$/gallon	$/liter	$/gallon
Sugarcane cost	0.22	0.84	0.25	0.95
Administrative and processing costs	0.10	0.38	0.12	0.47
Capital and other costs			0.12	0.45
Total cost	0.32	1.22[b]	0.49	1.87

Source: Ribera et al. 2007
[a]Chaves 2006
[b]Excludes capital costs.

The feedstock cost of $17/ton of cane paid to producers assumes that the ethanol plant covers the harvesting cost. Due to the U.S. sugar price support program, a ton of cane for sugar production in the U.S. is worth around $54, thus making sugarcane-based ethanol unable to compete with sugar production. Only molasses (juice from later pressings of sugarcane) is likely to prove anything close to economically feasible in the current policy environment.

The U.S. numbers should be viewed with some care as there is currently no sugarcane-based ethanol in the United States. There are relatively few other estimates for the cost of production with sugarcane-based ethanol. A USDA/LSU study showed the lack of economic feasibility to convert raw and refined sugar into ethanol in the U.S. (Shapouri et al. 2006). However, the costs of production cited above to convert sugarcane juice and/or molasses into ethanol, not raw and/or refined sugar. Moreover, a study by Rahmani and Hodges (2006) showed the cost of sugarcane-based ethanol ranging from $2.00 to $2.56 per gallon with a feedstock cost of $30 to $35 per ton of sugarcane.

Government Programs: *Sugar*

There are three primary U.S. sugar policy instruments in place. First, a price support loan program makes non-recourse loans to sugar processors (cf., marketing loans made to producers of other commodities) against sugar that is pledged as collateral. The loan rate is currently 18.5 cents per pound for raw cane sugar, and 23.77 cents per pound for refined beet sugar. Second, a marketing allotment program limits the quantity of sugar that marketers are allowed to sell each year. The government sets the overall quantity of sugar that can be marketed with the objective of avoiding forfeitures under the price support loan program. The overall allotment quantity is allocated amongst individual beet and cane processors using complex formulae. Third, a tariff rate quota (TRQ) program sets stiff tariffs for sugar imports above specific thresholds for each individual trading partner. Refined sugar TRQs are available predominantly in Canada and Mexico, while raw sugar TRQs are allocated among 40 countries based on trade patterns observed in the late 1970s and early 1980s. The aggregate TRQ is approximately one-eighth of U.S. sugar consumption. In-quota imports are subject to a tariff of less than one cent, while above-quota imports are subject to tariffs of 15.36 cents per pound of raw sugar and 16.21 cents per pound of refined sugar.

The net effect of the marketing allotment and TRQ programs is to limit total quantities of sugar sold in the U.S. and ensure that this sugar is mostly from U.S. producers. This raises prices well above free market levels; thus U.S. and world sugar prices are significantly disparate (Fig. 1). Discontent with U.S. sugar policy grows among policy analysts and policy makers (Groombridge 2001; Peterson 2011; Pitts and Davis 2011; Puente 2011). Sugar policy reform is the subject of at least one resolution introduced in the current legislative session (H.R. 1385, the Free Market Sugar Act). Criticisms leveled against U.S. sugar policy instruments include their amber box status under WTO rules (and that they consequently hinder trade negotiations and trade liberalization), that artificially high domestic sugar prices hurt U.S. consumers, and that U.S. import barriers hurt foreign agricultural workers.

Various studies in recent years have considered the partial equilibrium effects of sugar policy reform. Beghin et al. (2006) analyzed the effects of removing current U.S. sugar policy instruments, finding that U.S. sugar policy costs consumers $1.9 billion per year, and benefits U.S. producers and processors by over $1 billion. Deadweight loss is estimated to be over a half billion dollars per year. Elobeid and Beghin (2006) considered the effects of sugar policy reform in OECD countries more generally, finding that worldwide trade and production patterns would change substantially. Beghin (2006) and Abler et al. (2008) contended that due to changing WTO rules and mounting pressure from increasing trade under regional trade agreements, reform of U.S. sugar policy is essentially inevitable.

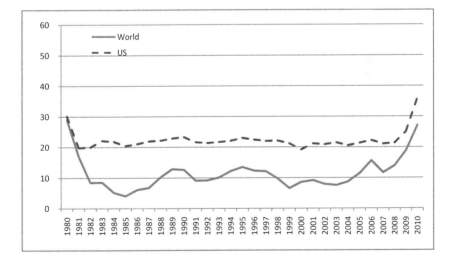

Figure 1. U.S. and World Raw Sugar Prices, 1980–2010 (calendar year average prices, in cents per pound shown on the y-axis, source: USDA-ERS).

Government Programs: *Ethanol*

Growth in the U.S. ethanol industry is directly related to Federal and State policies and regulations (Shapouri et al. 2006). Government incentives such as the motor fuel excise tax credits, small ethanol producer tax credits, import duties on fuel ethanol imports and others helped increase the production of ethanol during the 1980s. Government regulations, such the Clean Air Act Amendments of 1990, the Energy Policy Act of 1992, and the Energy Conservation Reauthorization Act of 1998, significantly increased the demand for ethanol during the 1990s. In recent years, the phasing out of MTBE, the Farm Security and Rural Investment Act of 2002, and the Energy Policy Act of 2005 along with surging prices for gasoline have sharply expanded the production and use of ethanol. It took 20 years for the ethanol industry to reach 1.6 billion gallons of production in 2000, but it took only 10 more years for the industry to increase ethanol production by almost ten-fold to over 13 billion gallons in 2010.

The Energy Policy Act of 2005 established the renewable fuels standard (RFS), which directs that gasoline sold in the U.S. should contain specified minimum volumes of renewable fuel. Under the Act, the total volume of renewable fuel to be utilized starts at 4 billion gallons in 2006 and increases to 7.5 billion gallons in 2012. However, most industry observers realize the Renewable Fuels Standard (RFS) contained in the Energy Policy Act of 2005 was never binding (Anderson et al. 2008). The Energy Independence and Security Act of 2007 amended the RFS signed into law in 2005 and increased the RFS, known as RFS2, to 36 billion gallons by 2022. Among other provisions, the RFS2 sets mandatory blend levels for renewable fuels while also establishing greenhouse gas (GHG) reduction criteria and methodology for calculating lifecycle GHG emissions. Therefore, the RFS2 limits the amount of grain-based ethanol to 15 billion gallons and the remaining 21 billion gallons must come from cellulosic and advanced biofuels. Sugarcane based ethanol is considered an advanced biofuel, which is to say it provides a favorable GHG reduction relative to grain-based biofuels, where the reductions are vis-a-vis comparable to fossil fuels (EPA 2010a,b).

There are several federal and state tax credits and incentives for the production, blending and/or sale of ethanol and ethanol blends. Two of the most debated about ones are the Volumetric Ethanol Excise Tax Credit (VEETC) and a tariff on imported ethanol. The VEETC, also known as the "blender's credit", is the primary federal tax incentive for the use of ethanol. This tax credit, which was created by the American Jobs Creation Act of 2004, provides blenders and marketers of fuel with a federal tax credit of 45 cents on each gallon of ethanol blended with their gasoline

(RFA). Through a market-based approach, VEETC enhances the sustained, cost competitiveness of ethanol with gasoline, and provides long-term protection against a volatile petroleum fuel market. As such, VEETC has been a major factor behind the spectacular increase in ethanol use, production and continued innovation in the industry. Because VEETC does not distinguish between feedstocks, it provides a tax benefit for all types of ethanol, including ethanol produced from new cellulosic and other second generation feedstocks, such as sorghum, sugarcane, energy cane, switchgrass, etc. Therefore, VEETC provides market-based, demand enhancement for not only the more traditional corn- and sugarcane-based ethanol, but also new and innovative types of ethanol that are becoming more and more commercially available. The VEETC expired on December 31, 2011. Moreover, due to the fact that all ethanol was eligible for the credit, to the extent that foreign ethanol received a tax benefit from VEETC, a secondary tariff was imposed on imported ethanol to offset that benefit (RFA). The secondary tariff was set at 54 cents per gallon. However, Congress has let both the VEETC and the secondary import tariff expire following arduous debates (Farm Policy 2012).

Relationship with the Broader Energy Market

In order to determine whether ethanol plants will remain profitable in the future it is important to understand its relationship with the broader transportation fuel market. Figure 2 illustrates the strong positive relationship between gasoline and ethanol prices (all measured on the left axis) and the acquisition costs of crude oil (measured on the right axis). One phenomenon that quickly jumps out is the large increase in ethanol prices during the summer of 2005 that is attributed to the unanticipated phase out of MTBE that is used as a summer oxygenate. While the graph helps explain trends, a more meaningful analysis is needed to see the actual statistical relationship between prices. The following simple equations were estimated using data from 2000 through 2010 to provide more insight into the price relationships, but not to predict or forecast fuel prices because these relationships may not hold in the future.

1) Monthly Ave. U.S. Price of Gasoline in $/Gal. = $0.7243 + 0.0259 * Price of Crude Oil/Barrel

2) Monthly Ave. U.S. Price of Ethanol in $/Gal. = $1.0037 + 0.0149 * Price of Crude Oil/Barrel

3) Monthly Ave. World Price of Raw Sugar in Cts./lb. = 4.7692 + 0.1476 * Price of Crude Oil/Barrel

4) Monthly Ave. World Price of Raw Sugar in Cts./lb. = 0.4164 + 6.8090 * Price of Ethanol/Gal.

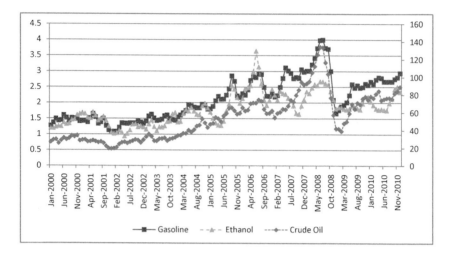

Figure 2. U.S. Prices of Crude Oil ($/barrel, right axis); Regular Gasoline and Ethanol ($/gallon, left axes); Monthly, January 2000–December 2010.

The R^2 goodness of fit measures for the U.S. gasoline equation was 0.95. The R^2 for the estimated ethanol equation was 0.60, which is to say that roughly 60 percent of the variability in ethanol prices can be explained by the variability in crude oil prices. This indicates that there are other factors such as government policies (i.e., the Renewable Fuel Standard and tax credits) affecting ethanol prices other than its role as a gasoline extender. In particular, during the period of low petroleum prices in the late 2008 and early 2009, U.S. ethanol volumes approached the RFS mandate levels leading to a temporary decoupling of U.S. ethanol prices and petroleum prices. There is a very low R^2 for world raw sugar and crude oil prices, 0.38, which shows that there is a low relationship between sugar prices and oil prices. In addition, there is an even lower relationship between world raw sugar prices and U.S. ethanol prices, 0.30. This minimal correspondence reflects the protected nature of the U.S. sugar market and the fact that ethanol consumed in the U.S. is largely domestically produced. This correspondence will be due only to the independent effects on each by changes in crude oil prices.

These relationships presented above will not hold with precision, but give an impression of the price relationships in recent years. The expiration of VEETC and the U.S. ethanol import tariff will have at least two effects on the relationships presented above. First, U.S. ethanol prices should command less of a premium above petroleum prices in the future. Second, the expirations will likely lead to increasing satisfaction of RFS2 by using imported ethanol, and an increasingly strong relationship between U.S. fuel prices and world sugar prices. In the absence of a major overhaul of U.S. sugar policy, however, U.S. sugarcane ethanol production is unlikely, and U.S. sugar prices will likely remain only loosely associated with energy prices.

References

Abler, D., J. Beghin, D. Blandford and A. Elobeid. 2008. Changing the U.S. sugar program into a standard crop program: consequences under the North America Free Trade Agreement and Doha. Review of Agricultural Economics 30: 82–102.

Anderson, D.P., J.L. Outlaw, H.L. Bryant, J.W. Richardson, D.P. Enrstes, J.M. Raulston, J.M. Welch, G.M. Knapek, B.K. Herbst and M.S. Allison. 2008. The Effects of Ethanol on Texas Food and Feed. Departmental working paper. Agricultural and Food Policy Center, Department of Agricultural Economics, Texas A&M University, College Station TX, April.

Beghin, J. 2006. A primer on US sugar in the 2007 US farm bill. Working paper, Center for Agriculture and Rural Development, University of Iowa, Ames, Iowa.

Chaves, I. 2006. Telephone conversation with the authors, 3 August.

Beghin, J., B. el Osta, J.R. Cherlow and S. Mohanty. 2006. The cost of the US sugar program revisited. Contemporary Economic Policy 21: 106–116.

Eggleston, G. 2010. Sustainability of the sugar and sugar-ethanol industries, chapter 1, pages 1–19. American Chemical Society, Washington, D.C.

Elobeid, A. and J. Beghin. 2006. Multilateral trade and agricultural policy reforms in sugar markets. Journal of Agricultural Economics 57: 23–48.

EPA. 2010a. EPA analyzes regulations for the national Renewable Fuel Standard program for 2010 and beyond. US Environmental Protection Agency Regulatory Announcement EPA-420-F-10-007.

EPA. 2010b. EPA lifecylce analysis of greenhouse gas emissions from renewable fuels. U.S. Environmental Protection Agency Regulatory Announcement EPA-420-F-10-006.

Farm Policy News. 2012. Biofuels Update. A summary of farm policy news. Available at: <http://farmpolicy.com/2012/01/13/budget-ag-economy-policy-issues-and-biofuels/> Accessed 13 January.

Groombridge, M.A. 2001. America's bittersweet sugar policy. Trade Brei_ng Paper 13, Cato Institute, Washington, D.C.

LSU Ag Center. 2012. 2012 Projected Louisiana Sugarcane Production Costs. Louisiana State University, Baton Rouge, Louisiana.

Matsuoka, S., J. Ferro and P. Arruda. 2009. The Brazillian experience of sugarcane ethanol industry. *In Vitro* Cellular & Developmental Biology—Plant 45(3): 372–281.

Peterson, M. 2011. American sugar policy leaves a sour taste. Web site http://www.policyinnovations.org/ideas/commentary/data/000136 viewed on December 5, 2011.

Pitts, J. and D. Davis. 2011. U.S. sugar policy: the real scary story. Web site http://thehill.com/blogs/congress-blog/economy-a-budget/190267-us-sugar-policy-the-real-scary-story, viewed on December 5, 2011.

Puente, M. 2011. Lugar says US sugar policy anything but sweet. Web site http://www.wbez.org/story/lugar-says-us-sugar-policy-anything-sweet-91282, viewed on December 5, 2011.

Rahmani, M. and A. Hodges. 2006. Potential Feedstock Sources for Ethanol Production in Florida. Food and Resource Economics Department, Florida Cooperative Extension Service, Institute of Food and Agricultural Sciences, University of Florida, Gainesville, Florida, August.

Renewable Fuels Association. 2012. Industry Statistics. Available at < http://www.ethanolrfa. org/pages/statistics>Accessed 05 December 2012.

Renewable Fuels Association (RFA). 2012. Tax Incentives. Available at <http://www.ethanolrfa. org/pages/tax-incentives>Accessed 06 December 2012.

Ribera, L.A., J.L. Outlaw, J.W. Richardson, J. da Silva and H. Bryant. 2007. Integrating Ethanol Production into a U.S. Sugarcane Mill: A Risk Based Feasibility Analysis. Departmental working paper. Agricultural and Food Policy Center, Department of Agricultural Economics, Texas A&M University, College Station TX.

Shapouri, Hossein, Michael Salassi and J. Nelson Fairbanks. 2006. The Economic Feasibility of Ethanol Production from Sugar in the United States. U.S. Department of Agriculture, July.

Texas AgriLife Extension Service. 2012. 2012 Texas Crop Enterprise Budgets. Texas A&M University System, College Station, Texas.

U.S. Department of Agriculture, Economic Research Service (ERS). 2011. US and World Sugar Prices, USDA-NASS, Washington DC.

U.S. Department of Agriculture, National Agricultural Statistics Service (NASS). 2011. 2010 Agricultural Statistics. USDA-NASS, Washington DC.

Vander Griend, D. 2006. The Impact of the Fuel Ethanol Industry on Rural America. Paper presented at the Farm Foundation Round Table meeting titled, Energy from Agriculture—Exploring the Future. Wichita, Kansas, June.

Sugarcane Genetic Transformation—Advances and Perspectives

André Luiz Barboza and *Helaine Carrer**

ABSTRACT

As a crop plant, sugarcane presents relatively narrow genetic diversity, low fertility and a complex genome. It is being considered as an important field crop for food and energy, supplying 80% of raw sugar production in the world, and it is currently used as bioenergy feedstock on a large scale in Brazil. Together, the three countries Brazil, India, and China are the major producers that generate 65% of the world production. As a renewable alternative to fossil fuel, worldwide demand for sugar and biomass has increased interest in adopting sugarcane for production agriculture worldwide. However, to realize the great potential of sugarcane as a bioenergy feedstock, improvements in sucrose accumulation, drought tolerance, diseases and insect resistance are urgently needed to increase the net productivity of this crop. To complement traditional breeding approaches, the

Department of Biological Sciences, Escola Superior de Agricultura "Luiz de Queiroz", Universidade de São Paulo, Av. Pádua Dias, 11. Piracicaba-SP. 13418-900. Brazil.
* Corresponding author

biotechnological tools of genetic transformation of sugarcane via biolistic and *Agrobacterium*-mediated gene transfer can accelerate introduction of desirable traits. This chapter will review the progress in establishing reliable gene transfer methods with sugarcane and summarize the progress to date on demonstrating the ability of this approach to product agronomically-relevant traits in this crop species. While ample examples of success for this technology are evident, current challenges and potential solutions to further improve this approach are discussed. Together with better agronomic practices and policies, the more rapid generation of elite sugarcane varieties with novel traits should lead to improved yield, additional bioproducts of commercial interest and to reduce undesirable environmental impacts.

Introduction

Sugarcane (*Saccharum officinarum*) is an important field crop for food and energy in tropical and subtropical regions of the world. It accumulates large amounts of sucrose in the stem that are used for the production of sugar and biofuel such as ethanol. Brazil is the largest of the producer countries followed by India and China. Together, the three countries generate up to 65% of the world production (FAOSTAT 2013). Sugarcane supplies over 80% of raw sugar production in the world and is also increasingly used as a bioenergy feedstock (http://www.fao.org). Ethanol production from sugarcane is considered to be one of the most viable alternatives to fossil fuels, not only to reduce dependence on crude oil but also to ameliorate CO_2 emissions generated from burning fossil fuels which contributes to global warming and climate change (de Souza et al. 2013). With the growing demand in sugar and biomass worldwide, there is an urgent need for biotechnological tools that can facilitate breeding programs towards developing new sugarcane cultivars with higher yield.

Sugarcane breeding using conventional approaches, in addition to better agronomical practices and policies, have contributed significantly to increase yields over the years. Selections for desirable traits such as disease and insect resistance, high sugar accumulation and easier harvesting have produced hundreds of commercial cultivars. In spite of the impressive achievements, however, classical breeding of sugarcane is a lengthy process due to a narrow genetic basis, low fertility and a complex genome. Thus, it takes normally at least 10 years of evaluation to release one or two commercial cultivars from hundreds of thousands of F1 seedlings (Arruda 2012). This long process limits the offer of new cultivars to overcome unexpected agronomic issues, which could bring significant reduction in yield or delayed achievement of industry targets such as second generation bioethanol and bio-based chemicals. Aiming to contribute to improvements in sugarcane productivity, genetic engineering technology of sugarcane could be a powerful tool to accelerate incorporation of new agronomic features and industrial strategies to face the challenges ahead and to reduce undesirable environmental impact.

Successful genetic engineering requires a reliable tissue culture and plant regeneration system, as well as an efficient transformation method. Plant regeneration through somatic embryogenesis was first used to successfully transform sugarcane via the biolistic approach by Bower and Birch (1992) and since then, genes for several agronomical traits such as herbicide tolerance, insect and pathogen resistance, increase of sugar accumulation, drought tolerance and high-value products such as PHAs have been introduced into sugarcane (Lakshmanan et al. 2005; Dal-Bianco et al. 2011; Arruda 2012). Field trials are also being conducted in several countries, although no commercial transgenic sugarcane cultivar exists at this time.

This chapter will highlight some of the biotechnological advances and present our perspectives for improving sugarcane in the near future using genomics and genetic engineering tools.

The Sugarcane Genome

Sugarcane is in the Poaceae family, and is a monocotyledonous grass belonging to the *Saccharum* genus. The modern commercial sugarcane cultivars are hybrids between *Saccharum officinarum* and *S. spontaneum* that exhibit high sugar content as well as the traits of disease and stress resistance (D'Hont et al. 1996). Due to interspecific crosses during evolution of the species, the genome constitution became more complex with chromosome number between 2n = 100–130 having been observed in different varieties. The chromosomal organization of the hybrids makes each cross a unique event due to the random sorting of chromosomes from both species affecting the distribution of favorable and unfavorable alleles (Grivet et al. 2001). In addition, the multiplicity of alleles (10 to 12 copies), makes the breeding program very complicated and long, taking on an average about 12 years to complete.

Regarding the complexity of the genome of modern sugarcane varieties, the use of molecular markers is essential to improve identification of allelic variations of individual cultivars (Lakshmanan et al. 2005). Data from DNA sequencing have contributed to the identification of molecular markers for desirable traits such as disease resistance, abiotic and biotic stress tolerance, with the interest to select new genotypes as a way toward generating new sugarcane varieties.

Initiatives of complete genome sequencing by research groups in Australia, France, South Africa, USA and Brazil is underway (http://sugarcanegenome.org). Already, 317 bacterial artificial chromosomes (BACs) were sequenced and generated 1,400 coding regions (De Setta et al. 2014). These data, in addition to sequences already deposited at the NCBI (http://www.ncbi.nlm.nih.gov), ESTs from the SUCEST and RNAseq projects (Tomkins et al. 1999; Vettore et al. 2003), together provide a remarkable source of genes and markers to be explored by molecular genetics tools. In this regard, transgenic introduction of relevant metabolic and agronomic genes into this highly complex genome can circumvent the difficulties of low fertility and long timetable that continue to plaque classical breeding approaches.

Sugarcane Tissue Culture

An efficient protocol for *in vitro* plant tissue culture is essential for genetic transformation. The first sugarcane tissue culture was reported by Nickel in 1964 showing callus formation and then plant regeneration was successfully obtained a few years later (Barba and Nickell 1969; Heinz and Mee 1969).

The most common explant used for callus formation and plant regeneration is immature leaves isolated from 4 to 8 month-old field grown sugarcane (Fig. 1A). It is obtained from the 10-centimeter long portion just above the apical meristem (Fig. 1B). After surface disinfection using 70% ethanol, and under aseptic conditions, two or three external leaves are removed. The remaining immature leaf roll formed by 5–6 furled leaves is sliced transversally into 1 to 2 mm thick sections that are placed on callus induction medium (Fig. 1D). Murashige and Skoog (MS) basal medium (Murashige and Skoog 1962) supplemented with 2,4-dichlorophenoxy acetic acid (2,4-D), sugars, vitamins, and coconut water (Figs. 1C, 1D) is commonly used to induce nodular embryogenic calli (Fig. 1C), which is the most preferred tissue for genetic transformation using either particle bombardment (biolistic) (Fig. 1E) or *Agrobacterium*-mediated transformation (Fig. 1F).

Genetic Transformation of Sugarcane

Stable plant transformation system consists of the introduction of a gene and all of the regulatory sequences into the genome of a totipotent plant

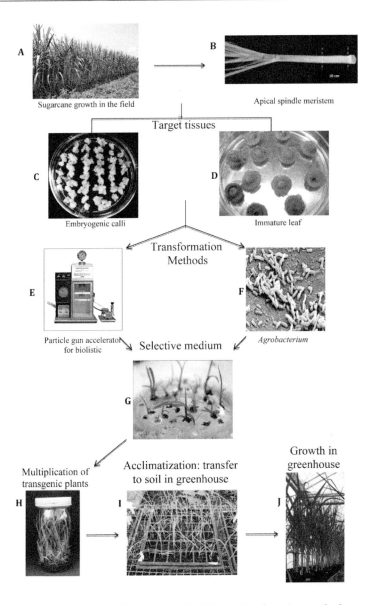

Figure 1. Steps of genetic transformation by biolistic and *Agrobacterium* methods to produce transgenic sugarcane plants. (A) sugarcane plantation in field as source of stem apices; (B) source of immature leaves; (C) Embryogenic calli; (D) Immature leaf rolls; (E) Particle Bombardment device for biolistic transformation; (F) *Agrobacterium tumefasciens* cells for transformation; (G) Putative transgenic plants on selective medium; (H) Transgenic sugarcane *in vitro* rooting and shoot multiplication; (I) Acclimatization in soil (greenhouse); (J) Growth in pots (greenhouse).

cell, followed by plant regeneration carrying the inserted elements and their inheritance in the progeny. Advances in recombinant DNA technology have allowed the insertion of foreign genes in diverse plant species of agronomic value. There are many variations of plant transformation methods to introduce and integrate the desirable gene(s) into the plant genome. The most widely used methods are *Agrobacterium*-mediated gene transfer, and transformation mediated by particle bombardment designated as biolistic methodology. Both methods require careful preparation of plant cells or tissues to receive the foreign DNA, selection of transformed cell lines, and efficient plant regeneration system. Only few methods do not require tissue culture for recovery of transgenic plants. They are commonly used in species for which efficient tissue culture are not available. The floral dip technique used for *Arabidopsis thaliana* (Clough and Bent 1998) and wheat (Zale et al. 2009) is a simplified and successful method. However, this approach could not be applied to most of the other plant species such as sugarcane, due to the low fertility of seeds and/or large plant size. Recently, sugarcane seeds were reported to be successfully transformed via *Agrobacterium tumefaciens* without tissue culture regeneration (Mayavan et al. 2013). Methods of *Agrobacterium* delivery using vacuum infiltration, sonication and bombardment of meristematic tissue in shoot apices and seeds are available. They can be simple, reduce need of highly sterile conditions, minimize somaclonal variations associated with tissue culture and decrease hands-on-time. However, many transformed shoots could be tissue chimeras that are not stable during their propagation.

The first transgenic sugarcane cells were obtained by PEG-mediated DNA transfer into protoplasts (Chen et al. 1987), this method is considered simple and does not require specialized equipment. However, it presented very low efficiency and reproducibility. For the same reasons, sugarcane transformation by electroporation (Rathus and Birch 1992) has not been used routinely. Considerable research efforts have been dedicated to develop efficient genetic transformation methodologies for sugarcane (Lakshmanan et al. 2005), demonstrating that *Agrobacterium*-mediated genetic transformation and the particle bombardment gene transfer method (biolistic) are the most successful approaches to achieve stable transgenic sugarcane plants.

Microprojectile Mediated Genetic Transformation (Biolistic method)

High-velocity microprojectiles used to deliver exogenous genetic material into plant tissue was developed and first described by John Sanford in collaboration with Ted Klein, Ed Wolf, and Nelson Allen (Sanford et al.

1987). This approach was also designated as biolistic (combination of "biological" and "ballistics") by the inventors. The first stable transformed plants obtained by biolistics were soybean (Christou et al. 1988) and tobacco (Klein et al. 1988). Because it is a direct gene transfer method, which does not rely on a biological vector, biolistics has been successfully used to transform cells that were recalcitrant to transformation by other means like some monocotyledonous species (Finer et al. 1999), and organelles such as plastids (Boyton et al. 1988; Svab and Maliga 1993) and mitochondria (Johnston et al. 1988).

In Australia, investigations on microprojectile-mediated transformation led to the development of the first transgenic sugarcane plant from a commercial cultivar by Robert Bower and Robert Birch in 1992. Since then, this approach has been used to transform sugarcane worldwide (Falco and Silva-Filho 2003; Altpeter and Oraby 2010). A list of sugarcane transformation events using biolistic is shown in Table 1.

The procedures to transform sugarcane with desirable genes follow some precise steps. The target tissue, either sugarcane embryogenic callus (Fig. 1C) or immature leaves (Fig. 1D), are usually used for transformation by abiolistic device (Fig. 1E). Gold or tungsten particles are coated with the vector DNA and are accelerated by a burst of inert gas (usually helium) under high-pressure in a vacuum chamber, and launched to the target tissue (Sanford 2000). After the bombardment, cell tissues are placed on MS basal medium in a dark chamber for three days to recover from the stress of physical impact before being transferred to regeneration medium supplied with an appropriate selectable agent, such as antibiotic or herbicide (Table 1), to select for and regenerate the putative transgenic plants (Fig. 1G). After positive identification of foreign DNA integration into the sugarcane genome by molecular analyses (polymerase chain reaction (PCR) or Southern blot), the transgenic plants are *in vitro* multiplied (Fig. 1H), transferred to a greenhouse in small vials for acclimatization (Fig. 1I) and to larger pots (Fig. 1J) later on for further genetic and physiological analyses.

The applicability to a wide range of target tissues and genotypes, and the simplicity of operation, make the microprojectile-mediated approach a very attractive and efficient method for sugarcane transformation.

Agrobacterium-mediated Genetic Transformation

Agrobacterium tumefaciens is a gram-negative soil bacterium responsible for causing crown gall tumors on a wide range of plants (De Cleene and De Ley 1976). The presence of a large tumor–inducing (Ti) plasmid in virulent strains of *Agrobacterium* is responsible for the transformed phenotype. This

Table 1. Traits engineered in transgenic sugarcane.

Method	P/Target gene/T	P/Selective marker/T	Reference
Abiotic stress tolerance			
Biolistic	Rab17/DREB2A CA/nos	NT/bar-PAT/NS	Reis et al. 2014
Biolistic	pS/AtCBL9-AtCIPK23-AtAKT1/nos	CaMV35S/hyg/polyA	Li et al. 2014
Bio-products			
Biolistic	NS/phaA, phaB, and phaC/NS	NS	Brumbley et al. 2008
Biolistic	Ubi-1/hchl and cpl/nos	Ubi-1/hptII/nos	McQualter et al. 2004b
Herbicide resistance			
Agrobacterium	Ubi-1/bar/nos	Ubi-1/bar/nos	Gallo-Meagher and Irvine 1996
Agrobacterium	CaM35S/bar/ocs	CaMV35S/gfp/ocs	Elliott et al. 1998
Agrobacterium	Ubi-1/bar/nos	ocs-CaMV35S/uidA-1/nos	Enríquez-Obregón et al. 1998
Biolistic	Ubi-1/bar/nos	Ubi-1/hptII/nos	Falco et al. 2000
Biolistic	Ubi-1/pat/nos	NS	Leibbrandt and Snyman 2003
Agrobacterium	CaMV35S/bar/nos	CaMV35S/nptII/uidA/nos	Manickavasagam et al. 2004
Biolistic	Ubi-1/als/nos	NS	van der Vyver et al. 2013
Insectresistance			
Electroporation	CaMV35S/tcryIA(b)/nos	CaMV35S/uidA/nos	Arencibia et al. 1998
Electroporation	CaMV35S/tcryIA(b)/nos	NS	Arencibia and Carmona 1999
Biolistic	Ubi-1/SBBI/nos Ubi-1/SKTI/nos	Ubi-1/nptII/nos	Falcoand Silva-Filho 2003
Biolistic	Ubi-1/s-cry1Ac/nos	Enu/nptII/nos	Weng et al. 2006
Agrobacterium	CaMV35S/Cry1Aa3/ocs	nos/nptII/nos	Kalunke et al. 2009

Biolistic	*Ubi-1/Cry1Ab/nos*	*CaMV35S/bar/polyA*	Arvinth et al. 2010
Agrobacterium	*CaMV35S/Cry1Aa3/ocs*	*nos/nptII/nos*	Kalunke et al. 2009
Biolistic	*Ubi-1/m-cry1Ac/nos*	*Emu/aphA/T-nos*	Weng et al. 2011
Lignin biosynthesis			
Biolistic	*osC4H/AS-COMT/CaMV35S*	*Ubi-1/nptII/CaMV35S*	Jung et al. 2012
Sucrose metabolism			
Biolistic	*Ubi-1/suc2/nos and Ubi-1/Anti-SCINVm/nos*	*Ubi-1/nptII/nos*	Ma et al. 2000
Biolistic	*Ubi-1/NI/nos*	*Ubi-1/uidA/nos*	Botha et al. 2001
Biolistic	*Ubi-1/PPO/nos*	*Emu/nptII/nos*	Vickers et al. 2005
Biolistic	*Ubi-1/NTPP-68J SI/nos*	*Emu/nptII/nos*	Wu and Birch 2007
Biolistic	*Ubi-1/mds6pdh -zmglk/nos*	*Ubi-1/nptII/nos*	Chong et al. 2007
Biolistic	*CaMV35S-Ubi-1/NI/nos CaMV35S/NI/nos*	NS	Rossouw et al. 2007
Biolistic	*Ubi-1/PFP/nos*	*Emu/nptII/nos*	Groenewald and Botha 2008
Biolistic	*Ubi-1/TS/nos*	*Ubi/nptII/nos*	Harmeli and Birch 2011
Virus resistance			
Biolistic	*Ubi/bar/nos*	*Ubi-1/hut/nos* *Ubi-1/nptII/nos*	Ingelbrecht et al. 1999
Biolistic	*Ubi-1/FDVS9ORF1/nos*	*Ubi-1/nptII/nos*	Mcqualter et al. 2004a
Biolistic	*Ubi-eut/SCMV/NS*	*Ubi-eut/nptII/NT*	Gilbert et al. 2005
Biolistic	*Ubi-1/SCYLV/nos*	*Ubi-1/nptII/nos*	Gilbert et al. 2009

Table 1. contd....

Table 1. contd.

Table 1. Successful engineering of traits in transgenic sugarcane. The linear representations of construction vectors depict the promoter (**P**) and the terminator (**T**) used for transgene expression and they are shown as uppercase P and T letters after the transgene name as P-promoter/gene/T-terminator. When promoter or terminator elements are not present, these were termed as Not Shown (NS). The names and sequence order of constructs used are shown as in the referenced publications: **als**: *acetolactate synthase gene*; **AphA- 2**: *synthetic NPT-II*; **AS-COMT**: *antisensecaffeic acid O-methyltransferase*; **AtAKT1**: *A. thalianaAKT1_inward shaker-like potassium channel gene*; **AtCBL9**: *A. thaliana CBL_Calcineurin β-like protein gene*; **AtCIPK**: *A. thaliana CIPK_CBL-interacting protein kinase gene*; **bar**: gene that encodes phosphinothricin acetyl transferase that inactivates phosphinothricin broad-spectrum herbicides by acetylation; **CaMV35S**: cauliflower mosaicvirus 35S promoter; **DREB2**: DRE-binding protein 2A; **Emu**: strong monocot promoter; **FDVS**: Fijidisease virus gene; **gfp**: green fluorescent protein; **huf**: untranslatable form of the Sorghummosaic virus (SrMV) strain SCH-CP (sorghum mosaic potyvirus strain coat protein) gene; **m-cry1Ac**: increased the GC content of cry1Ac; **MC**: minimal expression cassette; **manA**: Escherichia coli phosphomannose isomerase gene; **mds6pdh**: Malusdomesticasorbitol-6-phosphatedehydrogenase gene; **nos**: nopalinesynthetase gene terminator; **NI**: neutral invertase; **nptII**: neomycin transferaseII gene conferring resistance to aminoglycoside antibiotics such as kanamycin and geneticin; **NS**: not show; **osu**: artificial promoter consisting multiple octopinesynthase enhancer elements; **ocs**: octopine synthase terminator; **NTPP**: N-terminalpro-peptide gene; **OsC4H**: Oryza sativacinnamate 4-hydroxylase (C4H) gene promoter (GenBank accession no. AC136224); **pat**: phosphinotricinacetyl transferase gene; **PFP**: Pyrophosphate: fructose 6-phosphate 1-phosphotransferase in glycolytic carbon flow, which could be rate limiting under conditions of high metabolic activity; **PPO**: POLYPHENOLOXIDASE gene; **ProDIR16**: DIRIGENT (SHDIR16) gene stem-regulated expression; **SCSB**: sugarcane stem borer (Diatraeasaccharalis F.); **s-cry1Ac**: modified the cry1Ac gene to produce synthetic cry1Ac (designated s-cry1Ac) by increasing the GC content of the coding region following the codon usage preference of sugarcane; **SBB1**: soybean Bowman–Birkinhibitor gene; **SCYLV**: a Poleovirus of the Luteoviridae family; **TE**: transient expression; **tcry1A(b)**: truncated cry1A(b); **TS**: Trehalose synthase; **Ubi-1**: Maize uniquitin-1 promoter; **uidA**: a bacterial reporter gene that encodes β-Glucuronidase (UIDA) enzyme; **zmglk**: Zymomonasmobilis glucokinase.

plasmid carries a segment of DNA called T-DNA (Transferred DNA) that is mobilized from the bacterium to the plant and is inserted into random sites of the nuclear genome (Chilton et al. 1977). The Ti plasmid T-DNA contains genes for production of novel metabolites called opines that serve as nutrients for the inciting bacteria, and genes for synthesis of cytokinin and auxin, phytohormones that cause the proliferation of plant cells to form the gall (Klee et al. 1987). The phytohormone biosynthesis genes of the T-DNA were removed and replaced by foreign DNA, thus preventing the uncontrolled growth of the recipient cells and opened a world of application towards genetically modified crops (von Montagu 2011).

Agrobacterium-mediated transformation is the most commonly used method for dicotyledonous species, while for monocotyledonous species efficient transformation by this approach was considered almost a dream until the middle of the 1990s. One of the assumptions about the difficulty to transform monocots by *Agrobacterium* was the lack of an apparent wound response in these types of plants in contrast to dicots (Potrykus 1990). Thus, the use of cells that were actively dividing or about to divide and capable of regenerating plants is essential according to Hiei et al. (1994) and Ishida et al. (1996). Therefore, the authors discussed that types and stages of the tissues infected, the concentration of the inoculum, tissue culture media, the type of the vectors, the *Agrobacterium strains*, the selection markers and selective agents, and the genotype of plants, were of critical importance.

Agrobacterium-mediated transformation of sugarcane (Fig. 1F) was first achieved by using embryogenic calli (Bower and Birch 1992). The protocol used by the authors recommend the use of young regenerable calli (Fig. 1C) as target explants, co-cultivation in the dark for three days, and before transferring to selective medium (Fig. 1G), the callus from co-cultivation should be washed with the antibiotic cefotaxime. The selected shoots are then multiplied *in vitro* (Fig. 1H), and the rooted plants are transferred to small vials (Fig. 1I) and grown in the greenhouse for acclimatization. Later, they are transferred to large pots (Fig. 1J) for physiological and biochemical analyses. Over the years, this protocol has been improved and used to transform elite varieties (Elliot et al. 1998; Joyce et al. 2010). In addition to calli, Manickavasagam et al. (2004) also used auxillary buds as target tissue for transformation by *A. tumefaciens* strain EHA 105. A complete list of traits introduced into sugarcane by *Agrobacterium*-mediated transformation is shown on Table 1.

The efficiency of sugarcane transformation can be closely correlated with the choice of the *Agrobacterium* strain type and the plant genotype. A number of strains such as EHA101 and LBA4404 (Arencibia et al. 1998; Anderson and Birch 2012), AGL0 (Elliott et al. 1998), EHA105 (Manickavasagam et al. 2004), and AGL1 (Anderson and Birch 2012; Jackson et al. 2013) have been used for sugarcane transformation. Joyce at al. (2010) tested four *Agrobacterium* strains (AGL0, AGL1, EHA105 and LBA4404) with some variables affecting T-DNA delivery such as culture media, method of *Agrobacterium* inoculation, selection condition and co-cultivation period, concluded that the selection system and co-cultivation medium were the most important factors determining the success of sugarcane transformation. It is recommended to test *Agrobacterium* strains for each sugarcane genotype of interest in order to obtain the best efficiency of transformation.

Gene Copy Number and Expression Stability

Expression of instability of the transgene has been attributed to gene silencing induced by insertion of multiple copies in transgenic sugarcane or difficulties to sustain transgene expression once integrated into the genome after successive generations (Lakshmanan et al. 2005). The modified plants usually show high activity in young transformed seedlings, but expression can be turned off in mature plants in the first, second or third ratoons (Birch et al. 2010). Transgene silencing is apparently more frequent when using biolistic approach. This could be correlated with the fact that the biolistic methods tend to introduce many copies of the transgene into the genome of the recipient plant cell (Arvinth et al. 2010). Silencing-mediated transgene instability also seems to have a greater effect when the constructs incorporate promoters from sugarcane genes (Mudge et al. 2009; Arruda 2012).

Aiming to reduce insertion copy number in sugarcane genome transformation procedures with the biolistic method, a minimal expression cassette was developed. This approach uses transforming DNA fragments that lack a vector backbone and contain only the promoter, the coding sequence and the 3' UTR sequence. It showed efficient integration and expression in sugarcane (Taparia et al. 2012; Joyce et al. 2014). Sugarcane transformation using a firefly luciferase reporter gene in a minimal expression cassette via biolistic method, showed many single-copy transgenic lines when compared to *Agrobacterium*-mediated gene transfer, suggesting that the minimal expression cassette approach could be used without reduction of transformation efficiency (Jackson et al. 2013).

Since expression of foreign genes in sugarcane is often limited by transgene silencing, silencing through hairpin-mediated RNA interference (RNAi) has been tested to down-regulate endogenous sugarcane genes in a highly polyploid and heterozygous genome. Using a hairpin-mediated silencing construct of the phytoene desaturase (*PDS*) gene, almost all the endogenous transcripts of PDS can be eliminated as shown by northern analysis complemented by observed photo-bleaching of chlorophyll in the leaves due to the PDS knock-down. These results indicate that hairpin-mediated down-regulation is an efficient approach in sugarcane in spite of the complex genome of this plant, thus showing its promise as a new tool for improvement of sugarcane (Osabe et al. 2009).

Promoters and Selective Markers for Transgene Expression

The establishment of an efficient protocol for creating transgenic sugarcane plants requires a selectable agent in the medium to identify the putative transgenic shoots, eliminating possible chimeric plants and escapees. In addition, successful recovery of transformed plants can occur when the selectable marker gene is placed under the control of a strong constitutive promoter (Joyce et al. 2010).

There are many selectable marker genes reported for use in generating transgenic sugarcane. The most commonly used are those that encode *neomycin phosphotransferase* II (*nptII*), which inactivates the antibiotic kanamycin by phosphorylation, and phosphinothricin acetyl-transferase (*bar*) that inactivates phosphinothricin herbicides by acetylation (Table 1). The use of kanamycin as selective agent for sugarcane is discouraged by its low selection efficiency due to high levels of endogenous resistance and to deleterious effects on regeneration of transformed cells of monocots (Hauptmann et al. 1988; Nehra et al. 1994). Thus, the analogue geneticin (G418) is frequently used as a selection agent with the *nptII* gene for sugarcane (Bower and Birch 1992; Jung et al. 2012). The other antibiotic that is commonly used is hygromycin phosphotransferase (*hph*) (Arencibia et al. 1998; Joyce et al. 2010). Also, phosphomannose isomerase (PMI), has been used for positive selection of transformed sugarcane tissues, which allows growth of the transformed cells and tissues on mannose-containing medium (Jain et al. 2007). The reporter genes most often used for sugarcane are currently β-glucuronidase (GUS) (Jefferson 1987; Liu et al. 2003) and green fluorescent protein (GFP) (Elliott et al. 1998).

Common promoters used in engineering sugarcane are the Cauliflower Mosaic virus 35S Promoter (*CaMV35S*), maize *ubiquitin*-1 (*Ubi*-1), rice-*Ubiquitin* (*RUBQ2*), rice actin (*Act*1) and the synthetic *Emu* (Christensen and Quail 1996; Liu et al. 2003; Lakshmanan et al. 2005). The *CaMV35S* promoter has been considered as a weak one for sugarcane expression when compared with the *Ubi*-1 promoter (Rathus et al. 1993; Gallo-Meagher and Irvine 1996). There are only a few promoters isolated and characterized from sugarcane (Table 1).

Availability of tissue-specific promoters is still a challenge in sugarcane. Strong promoters, for stem and root tissues, can be very valuable for production of bio-based compounds and for crop improvement. Accumulation of insect control proteins in the stem and root would benefit the control of pests. Also, root specific promoters could help drive research on efficient water use, nutrient uptake, and disease control in sugarcane.

Traits Introduced into Sugarcane

As a powerful tool that can rapidly incorporate genes to confer agronomically desirable traits such as drought tolerance, as well as disease, virus and insect resistance, genetic transformation of sugarcane can help remove limitations for the improvement of elite sugarcane cultivars. In addition to increasing stress tolerance, this technology also makes it possible to further improve elite cultivars through directed increase in the sucrose yield, or changing fibers production in mature sugar stalks, thus promoting overall biomass yield for energy production (Arruda 2012). Recent advances in transgenic sugarcane plants are summarized on Table 1 and described as follow:

Virus and Insect Resistance

Efforts have been made to develop transgenic sugarcane with improved resistance to microbial pathogens and insect pests. Transgenic sugarcane overexpressing virus coat proteins or antisense RNA to suppress virus proliferation, are approaches used to generate sugarcane plants resistant to sugarcane mosaic virus (SCMV; Joyce et al. 1998), sugarcane yellow leaf virus (SCYLV; Gilbert et al. 2009; Zhu et al. 2011), sorghum mosaic potyvirus (SrMV; Ingelbrecht et al. 1999), and Fiji disease virus (FDV; McQualter et al. 2004a). These modifications on the sugarcane genome improved virus resistance and can reduce costs of sugarcane production as observed in experimental fields (Gilbert et al. 2009).

Insect attack is one of the major causes of economic loss that impact negatively on sugarcane productivity. Transgenic sugarcanes expressing endotoxins produced by *Bacillus thuringiensis* (Bt) have been studied for protection to Lepdopteran insects of importance in sugarcane such as the stem borer (*Diatraea saccharalis* F.) and sugarcane giant borer (*Telchin licus licus* Drury). The *cryIA*(b) gene under control of CaMV35S promoter showed a low level of expression. Nevertheless, it was found to be sufficient to improve resistance to sugarcane borer neonate larvae (Arencibia et al. 1999). Higher levels of the Bt endotoxin proteins were obtained via optimized

Cry1A transgene codon usage along with the control by a maize ubiquitin promoter (Arvinth et al. 2010; Weng et al. 2011). The *Cry1Aa3* gene was used with *Agrobacterium*-mediated transformation on immature leaf discs to avoid somaclonal variation and improve transgene stability (Kalunke et al. 2009). Transgenic sugarcane carrying the *cry1A*(b) gene driven by the *Ubi*-1 maize promoter was retransformed to stack it with the *bovine pancreatic trypsin inhibitor* (*aprotinin*) gene. The bioassay of transgenic plants under attack by shoot borer larvae displayed a synergistic pattern of damages with low interference by aprotinin in Cry1A(b)'s activity (Arvinth et al. 2010). The introduction of *soybean Kunitz tryps in inhibitor* (*SKTI*) and *soybean Bowman–Birk inhibitor* (*SBBI*) genes in sugarcane resulted in low mortality and retardation of larval growth in laboratory bioassays. However, their expression did not minimize insect damage in greenhouse trials (Falco and Silva-Filho 2003).

Herbicide Resistance

Sugarcane herbicide resistance was obtained by introduction and expression of the *bar* gene with the biolistic (Gallo-Meagher and Irvine 1996; Ingelbrecht et al. 1999) and *Agrobacterium*-mediated transformation methods (Enriquéz-Obregón et al. 1997; Manickawasagan et al. 2004). Transgenic plants displayed the expected resistance to bialaphos and glufosinate ammonium herbicides. In contrast, plants transformed with the *pat* gene (Leibbrandt and Snyman 2003), which also encodes for phosphinothricin ammonium transferase did not show any morphological difference when compared to control plants. Currently, the *bar* gene is one of the most commonly used selection marker to obtain transgenic sugarcane (Ingelbrecht et al. 1999; Falco et al. 2000). Searching for new selection marker gene, the tobacco *acetolactate synthase* (*als*) gene driven by the *Ubi*-1 promoter was used to transform sugarcane by biolistic method. The transgenic events showed expression under sulfonylurea herbicide selection (van der Vyver et al. 2013). Besides the potential benefits of herbicide resistance plants to the environment by decreasing the use of herbicide sprays, these results introduce an alternative marker gene from plant origin that may facilitate regulatory approval of the transgenic material through avoiding the use of bacterial selection genes.

Sucrose Metabolism

Sugar accumulation is a focus for engineering sucrose metabolism in sugarcane for food and biofuel production. Sugarcane is a C4 plant fixing carbon in the mesophyll cells to be converted into soluble sugars by a complex photosynthesis process, which are then stored in parenchyma cells that are localized in the mature stalks (McCormick et al. 2009). In sugarcane mature stalks, sucrose is stored at concentrations of 600–700 mM in the acidic vacuoles, which occupy 80% of the parenchyma cells' volume and is limited by the osmotic gradient between the vacuoles and symplast (McCormick et al. 2009; Rae et al. 2009; Arruda 2012). Aiming to overcome this osmotic constraint of sucrose content in vacuoles, transgenic sugarcane engineered to down-regulate the levels of the pyrophosphate:fructose 6-phosphate 1-phosphotransferase (PFP) enzyme was produced, using constitutive overexpression of antisense RNA targeting PFP-b transcripts. A reduction of PFP levels was observed in immature stalks and an increase of sucrose up to eightfold can be detected in the immature internodes as well as higher fiber content due to elevated hexose-phosphate levels (Groenewald and Botha 2008; van der Merwe et al. 2010). Another approach to manipulate sugar metabolic pathway was through the introduction of a sucrose isomerase gene designed for vacuolar compartmentalization. Transgenic sugarcanes with this construct showed an increase of two-fold in the total sugar content, sustained production of this isomer in the vegetative progenies (Hamerli and Birch 2011), and at the same time increases in photosynthesis activity and sucrose transport were observed (Wu and Birch 2007). Transgenic sugarcane events expressing antisense orientation of *soluble acid invertase* cDNA (*SCINVm*) showed reduction in the soluble invertase activity resulting in an increase of sucrose content in the cells (Ma et al. 2000; Botha et al. 2001). However, when a sense *Saccharomyces cerevisiae invertase* gene (*SUC2*) was used, there was an increase of invertase activity and reduction of sucrose content, principally in the apoplast. Consequently, there was a decrease of the sucrose content in the stalks (Ma et al. 2000). Furthermore, when transgenic sugarcane was produced containing an antisense neutral invertase gene, changes in the conduction of the carbon toward respiratory process in the stalks was observed (Rossouw et al. 2007). Transgenic sugarcane lines expressing the sense or antisense orientation *polyphenoloxidase* (*PPO*) gene are currently being investigated. Manipulation of this metabolic enzyme-encoding gene resulted in alteration of PPO activity and reduction in color intensity of the juice and raw sugar, which demonstrates that the juice and sugar color are correlated with PPO activity (Vickers et al. 2005).

Biomass Production

A large part of the fixed carbon in sugarcane stalks is in the cellulose and hemicellulose forms (de Souza et al. 2013). The non-harvested portions of sugarcane, in the form of leaves and the leftover bagasse can be used as feedstock for second-generation ethanol production (Arruda 2012;

Buckeridge et al. 2012). However, currently this process to produce ethanol has a high cost of enzymes that are necessary to degrade cellulose and hemicellulose into fermentable sugars (Sainz 2009). To improve the economics of this process, transgenically-expressed cellobiohydrolases and endoglucanase were targeted to vacuole and chloroplast compartments of sugarcane cells, respectively (Harrison et al. 2011). The significant accumulation of these enzymes in the sugarcane plants shows a potential of this approach for improvement of second generation ethanol production with sugarcane biomass by using energy cane engineered with their own cellulolytic enzymes.

Modification of lignin biosynthesis and increase of biomass production are also goals for biofuel and plant-based feedstock improvements. Transgenic sugarcane obtained by using the RNAi minimal cassette as a strategy to insert the antisense *COMT* gene fragment into the genome of sugarcane resulted in a reduction of the lignin content and increase of the sucrose content stored in the stalks (Jung et al. 2012).

Biotic Stress

Drought is one of the major limitation factors that can significantly decrease yield. Sugarcane plants with the ability to tolerate longer periods of water deficit have been developed by the insertion of genes and transcription factors into the plant. Expression of the trehalose synthase gene (Zhang et al. 2006) and pyrroline-5-carboxylate synthetase (P5CS), an enzyme involved in the proline biosynthesis pathway, can improve water stress tolerance and the expression of the P5CS protein also correlates with increased biomass production when the transgenic plants are exposed to drought stress (Molinari et al. 2007). Transgenic sugarcane plants overexpressing *AtDREB2ACA* gene exhibit improved response to water stress and expressing several classes of genes that are involved in activating different response pathways against dehydration. Consequently, these plants maintained higher levels of relative water content (RWC) and water potential when compared to the control sugarcane plants after four days without supply of water. Furthermore, the drought tolerance of *AtDREB2A*-expressing sugarcane lines showed higher photosynthetic rates at the third day during water-deficit period without biomass loss (Reis et al. 2014).

Sugarcane has a high biomass production and consumes a large quantity of sunlight for optimal productivity, and nutrients that can be rapidly drained from the soil, such as potassium (K$^+$), the low availability of which can limit sugarcane growth and production. To improve the capacity for tolerance to low soil-potassium concentrations, three components of the particular stress signaling pathway, AtCBL9, AtCIPK23 and AtAKT1, from *Arabidopsis thaliana* were used to genetically transform sugarcane. Transgenic plants showed increases of potassium content in tissue culture and hydroponic culture conditions. Under low potassium availability conditions, these transgenic sugarcane displayed longer roots, higher plant height, heavier dry weight, thus demonstrating the ability to engineer increased tolerance to low-potassium stress, with potential improvement in sugarcane production for areas where the ground is poor in potassium (Li et al. 2014).

Bio-products

Sugarcane has all the features needed for a natural biofactory: it grows fast, has a very efficient carbon fixation pathway, produces a large amount of biomass, possesses a well-developed storage system (stem) with a large pool of hexose sugar, and is cultivated in different parts of the world (Lakshmanan et al. 2005). Industrial products such as poly-3-hydroxybutyrate (PHB) and p-hydroxy-benzoic acid (pHBA) used as biodegradable carbon (such as plastic) have been demonstrated to accumulate in transgenic sugarcane upon expression of genes encoding the enzymes of PHB production and chorismate pyruvate-lyase (CPL), respectively (McQualter et al. 2004b; Brumbley et al. 2008). The expression was higher in leaves, accumulating up to 7% of pHBA, while in the stem this product reached to only 1.5%. Similar results were obtained with PHB showing that molecular biology and plant physiology knowledge may be needed to direct the accumulation of this bio-product to the stem, the storage organ of sugarcane, in order to further maximize the quantity of the target products.

Perspectives

The progress made in sugarcane genetic transformation and biotechnology during the last two decades is remarkable but still there is a need of more advancements in order to achieve the enormous potential for enhancing sugarcane performance through the incorporation of desirable traits in elite cultivars.

The Expressed Sequence Tags (ESTs) of sugarcane (Vettore et al. 2003) opened the opportunity to use bioinformatic tools for discovery of sugarcane genes that may confer improvement on sugar yield in mature stalk, drought tolerance and disease resistance (Souza et al. 2011). Genomics has been shown to be successful for facilitating marker-assisted breeding programs to identify superior genotypes and efficient breeding strategies. Moreover, candidate genes can now be efficiently introduced into the genome to produce transgenic sugarcane, thus making possible more rapid improvement of sugarcane elite varieties for even higher yield and biomass production. Together with the increasing amount of transcriptome sequences that are available, the efforts of a reference genome will help speed up gene discovery projects and the identification of useful promoter sequences for high level and selective expression of desirable genes. Strong tissue-specific promoters, particularly for stem and root tissues, would be very valuable for biofactory and sugarcane improvement research (Lakshaman et al. 2005). Manipulation of product accumulation in the stem to control insect infestation by a transgenic approach will also depend in part on the availability of promoters that are specific for or highly active in stem or root tissue.

The major methods of sugarcane genetic transformation are the particle-bombardment (biolistic) and *Agrobacterium*-mediated transformation approaches. Although the efficiency of transformation is currently not high, they are ready for optimization through learning from successes achieved in major plant model systems that may be applicable to sugarcane. The level of expression and stability maintenance of the transgene in the plant are important constraints for genetically modified sugarcane to overcome and will be necessary to pave the road towards commercial release. While there are currently no commercial transgenic sugarcane varieties on the market, there are ongoing experimental field trails. Advances on specific promoters, selectable markers, the use of minimal cassettes and selection of *Agrobacterium* strains will offer new strategies for improving the efficiency of sugarcane engineering in the near future.

Genetic modification of sugarcane has now been shown to be a powerful tool for incorporating key agronomic traits such as disease resistance, insect resistance, drought tolerance and increase of sugar yield (Lakshman et al. 2005; Arruda 2012). Therefore, genetic transformation technology associated with metabolic engineering will become an important approach for biofuel and biomaterials development of this species. Sugarcane plants produce high amount of biomass, being the best feedstock for biofuel production with more favorable energy input/output ratio than corn, also another major biofuel feedstock (Lam et al. 2009; de Souza et al. 2013). Thus, sugarcane provides a large potential to be used in industrial approaches for biofuel and biomaterial production that are not dependent on petroleum, such as biodegradable plastic or intermediates for the chemical industry (von Montagu 2011).

To achieve high level expression of a desired transgene product, genetic engineering of the sugarcane plastid genome can potentially offer a way to introduce multi-gene cassettes in a single transformation event. In addition to the high copy number of plastids per cell, this approach also minimizes the possibility of unintended spread of the transgene due to maternal inheritance of the plastid genome, absence of gene silencing, and potential elimination of the selectable marker DNA. A large number of transgenes, conferring high expression of desirable agronomic and biopharming traits, have been integrated stably into the tobacco plastid genome (Maliga and Bock 2011; Rogalski and Carrer 2011). The expansion of the plastid transformation technology to agronomically relevant plants, such as sugarcane, is still a challenge since no monocots has had the plastid genome stably modified till now. Considering there is a protocol for sugarcane plant regeneration from direct somatic embryogenesis from immature leaves and that the sugarcane plastid genome is available (Asano et al. 2004; Calsa-Junior et al. 2004) for designing expression vectors, this approach can be explored in the near future to further develop additional avenues for engineering key traits in sugarcane.

References

Altpeter, F. and H. Oraby. 2010. Sugarcane. pp. 453–472. *In*: F. Kempken and C. Jung (eds.). Genetic Modification of Plants. Berlin, Springer.

Anderson, D.J. and R.G. Birch. 2012. Minimal Handling and Super-Binary Vectors Facilitate Efficient, *Agrobacterium*-Mediated, Transformation of Sugarcane (*Saccharum* spp. hybrid). Trop. Plant Biol. 5: 183–192.

Arencibia, A., E.R. Carmona, P. Tellez, M.T. Chan, S.M. Yu, L.E. Trujillo and P. Oramas. 1998. An efficient protocol for sugarcane (*Saccharum* spp. L.) transformation mediated by *Agrobacterium tumefaciens*. Transgenic Res. 7: 213–222.

Arencibia, A., E. Carmona, M.Y. Cornide, S. Castiglione, J. O'Relly, A. Cinea, P. Oramas and F. Sala. 1999. Somaclonal variation in insect-resistant transgenic sugarcane (*Saccharum* hybrid) plants produced by cell electroporation. Transgenic Res. 8: 349–360.

Arruda, P. 2012. Genetically modified sugarcane for bioenergy generation. Curr. Opin. Biotechnol. 23: 315–22.

Arvinth, S., S. Arun, R.K. Selvakesavan, J. Srikanth, N. Mukunthan, P.A. Kumar, M.N. Premachandran and N. Subramonian. 2010. Genetic transformation and pyramiding of aprotinin-expressing sugarcane with *cry1Ab* for shoot borer (*Chiloinfuscatellus*) resistance. Plant Cell Rep. 29: 383–395.

Asano, T., T. Tsuduki, S. Takahashi, H. Shimada and K.-I. Kadowaki. 2004. Complete Nucleotide Sequence of the Sugarcane (*Saccharum officinarum*) Chloroplast Genome: A comparative Analysis of Four Monocot Chloroplast Genomes. DNA Research 11: 93–99.

Barba, R. and L.G. Nickell. 1969. Nutrition and organ differentiation in tissue culture of sugarcane—a monocotyledon. Planta 89: 299–302.

Birch, R.G., R.S. Bower and A.R. Elliot. 2010. Highly efficient, 5-sequence-specific transgene silencing in a complex polyploidy. Top. Plant Biol. 3: 88–97.

Botha, F.C., B.J.B. Sawyer and R.G. Birch. 2001. Sucrose metabolism in the culm of transgenic sugarcane with reduced soluble acid invertase activity. *In*: D.M. Hogarth (ed.). Proc. Int. Soc. Sugar Cane Technol., Brisbane 24: 588–591.

Bower, R. and R.G. Birch. 1992. Transgenic sugarcane plants via micro-projectile bombardment. Plant Journal 2: 409–416.

Boyton, J.W., N.W. Gillham, E.H. Harris, J.P. Hosler, A.M. Johnson, A.R. Jones, B.L. Randolph-Anderson, D. Robertson, T.M. Klein, K.B. Shark and J.C. Sanford. 1988. Choroplast transformation in Chlamydomonas with high velocity microprojectiles. Science 240: 1524–1538.

Buckeridge, M.S., A.P. De Souza, R.A. Arundale, K.J. Anderson-Teixeira and E. DeLucia. 2012. Ethanol from sugarcane in Brazil: a 'midway' strategy for increasing ethanol production while maximizing environmental benefits. GCB Bioenergy 4: 119–126.

Brumbley, S.M., L.A. Petrasovits, P.A. Bonaventura, M.J. O'Shea, M.P. Purnell and L.K. Nielsen. 2008. Production of polyhydroxyalkanoates in sugarcane. Proc. Int. Soc. Sugar Cane Technol. Mol. Biol. Workshop, Montpellier, France 4: 31.

Calsa Junior, T., D.M. Carraro, M.R. Benatti, A.C. Barbosa, J.P. Kitajima and H. Carrer. 2004. Structural features and transcript-editing analysis of sugarcane (*Saccharum officinarum* L.) chloroplast genome. Curr. Genet. 46: 366–373.

Chen, W.H., K.M.A. Gartland, M.R. Davey, R. Sotak, J.S. Gartland, B.J. Mulligan, J.B. Power and E.C. Cocking. 1987. Transformation of sugarcane protoplasts by direct uptake of a selectable chimaeric gene. Plant Cell Rep. 6: 297–301.

Chilton, M.-D., M.H.D. Donald, J. Merlo, D. Sciaky, A.L. Montoya, M.P. Gordont and E.W. Nester. 1977. Stable Incorporation of Plasmid DNA into Higher Plant Cells: the Molecular Basis of Crown Gall Tumorigenesis. Cell 11: 263–271.

Chong, B.F., G.D. Bonnett, D. Glassop, Michael G. O'Shea and S.M. Brumbley. 2007. Growth and metabolism in sugarcane are altered by the creation of a new hexose-phosphate sink. Plant Biotechnol. J. 5: 240–253.

Christensen, A.H. and P.H. Quail. 1996. Ubiquitin promoter-based vectors for high-level expression of selectable and/or screenable marker genes in monocotyledonous plants. Transgenic Res. 5: 213–318.

Christou, P., D.E. McCabe and W.F. Swain. 1988. Stable transformation of soybean callus by DNA coated particles. Plant Physiol. 87: 671–674.

Clough, S.J. and A.F. Bent. 1998. Floral dip: A simplified method for *Agrobacterium*-mediated transformation of *Arabidopsis thaliana*. Plant J. 16: 735–743.

Dal-Bianco, M., M.S. Carneiro, C.T. Hotta, R.G. Chapola, H.P. Hoffmann, A.A. Garcia and G.M. Souza. 2011. Sugarcane improvement: how far can we go? Curr. Opin. Biotechnol. 23: 1–6.

De Cleene, M. and J. De Ley. 1976. The host range of crown gall. Botanical. Rev. 42: 389–466.

De Setta, N., C. Monteiro-Vitorello, C. Metcalfe, G. Cruz, L. Del Bem, R. Vicentini, F.T.S. Nogueira, R.A.C. ampos, S.L. Nunes, P.C.G. Turrini, A.P. Vieira, E.A.O. Cruz, T.C.S. Corrêa, C.T. Hotta, A. de M. Varani, S. Vautrin, A.S. da Trindade, M. de M. Vilela, C.G. Lembke, P.M. Sato, R.F. de Andrade, M.Y. Nishiyama, Jr., C.B. Cardoso-Silva, K.C. Scortecci, A.A.F. Garcia, M.S. Carneiro, C. Kim, A.H. Paterson, H. Bergès, A. D'Hont, A.P. de Souza, G.M. Souza, M. Vincentz, J.P. Kitajima and M.A. Van Sluys. 2014. Building the sugarcane genome for biotechnology and identifying evolutionary trends. BMC Genomics 15: 540.

de Souza, A.P., A. Grandis, D.C.C. Leite and M.S. Buckeridge. 2013. Sugarcane as a Bioenergy Source: History, Performance, and Perspectives for Second-Generation Bioethanol., Bio. Energy Research 7, 1, 24.

D'Hont, A., L. Grivet, P. Feldmann, S. Rao, N. Berding and J.C. Glaszmann. 1996. Characterization of the double genome structure of modern sugarcane cultivars (*Saccharum* spp.) by molecular cytogenetics. Molecular Gene Genet. 250: 405–413.

Elliott, A.R., J.A. Campbell, R.I.S. Bretell and C.P.L. Grof. 1998. *Agrobacterium*-mediated transformation of sugarcane using GFP as a screenable marker. Aust. J. Plant Physiol. 25: 739–743.

Enríquez-Obregón, G.A., R.I. Vázquez-Padrón, D.L. Prieto-Samsonov, G.A. De la Riva and G. Selman-Housein. 1997. Herbicide-resistant sugarcane (*Saccharum officinarum* L.) plants by *Agrobacterium*-mediated transformation. Planta 206: 20–27.

Falco, M.C., A.N. Tulmann and E.C. Ulian. 2000. Transformation and expression of a gene for herbicide resistance in Brazilian sugarcane. Plant Cell Reports 19: 1188–1194.

Falco, M.C. and M.C. Silva-Filho. 2003. Expression of soybean proteinase inhibitors in transgenic sugarcane plants: effects on natural defense against *Diatraesaccharalis*. Plant Physiol. Biochem. 41: 761–766.

FAOSTAT. 2013. Feeding the world. Fao satatistical yearbook 154. http://faostat3.fao.org/ last access, 14 december, 2014.

Finer, J.J., K.R. Finer and T. Ponappa. 1999. Particle Bombardment Mediated Transformation. pp. 59–80. *In*: J. Hammond, P. McGarvey and V. Yusibov (eds.). Plant Biotechnology. New Products and Applications. Springer-Verlag Berlin Heidelberg, Germany.

Gallo-Meagher and M.J.E. Irvine. 1996. Herbicide resistant transgenic sugarcane plants containing the *bar* gene. Crop Sci. 36: 1367–1374.

Gilbert, R.A., M. Gallo-Meagher, J.C. Comstock, J.D. Miller, M. Jain and A. Abouzid. 2005. Agronomic evaluation of sugarcane lines transformed for resistance to Sugarcane mosaic virus strain. E. Crop Sci. 45: 2060–2067.

Gilbert, R.A., N.C. Glynn, C. Comstock and M.J. Davis. 2009. Agronomic performance and genetic characterization of sugarcane transformed for resistance to sugarcane yellow leaf virus. Field Crops 111: 39–46.

Grivet, L., J.C. Glaszmann and P. Arruda. 2001. Sequence polymorphism from EST data in sugarcane: A fine analysis of 6-phosphogluconate dehydrogenase genes. Genet. Mol. Biol. 24: 161–167.

Groenewald, J.-H. and F.C. Botha. 2008. Down-regulation of pyrophosphate: fructose 6-phosphate 1-phosphotransferse (PFP) activity in sugarcane enhances sucrose accumulation in immature internodes. Transgenic Res. 17: 85–92.

Hamerli, D. and R.G. Birch. 2011. Transgenic expression of trehalulose synthase results in high concentrations of the sucrose isomer trehalulose in mature stems of field-grown sugarcane. Plant Biotechnol. J. 9: 32–37.

Harrison, M.D., J. Geijskes, H.D. Coleman, K. Shand, M. Kinkema, A. Palupe, R. Hassall, M. Sainz, R. Lloyd, S. Miles and J.L. Dale. 2011. Accumulation of recombinant cellobiohydrolase and endoglucanase in the leaves of mature transgenic sugar cane. Plant Biotechnol. J. 1: 1–13.

Hauptmann, R.M., V. Vasil, P. Ozaias-Aikins, Z. Tabaeizadeh, S.G. Rogers, R.T. Fraley, R.B. Horsch and I.K. Vasil. 1988. Evaluation of selectable markers for obtaining stable transformants in the Gramineae. Plant Physiol. 86: 602–606.

Heinz, D.I. and G.W.P. Mee. 1969. Plant differentiation from callus tissue of *Saccharum* species. Crop Sci. 9: 346–348.

Hiei, Y., S. Ohta, T. Komari and T. Kumashiro. 1994. Efficient transformation of rice (*Oryzasativa* L.) mediated by *Agrobacterium* and sequence analysis of the boundaries of the T-DNA. Plant J. 6: 271–282.

Ingelbrecht, I.L., J.E. Irvine and T.E. Mirkov. 1999. Posttranscriptional gene silencing in transgenic sugarcane. Dissection of homology-dependent virus resistance in a monocot that has a complex polyploid genome. Plant Physiol. 119: 1187–1197.

Ishida, Y., H. Saito, S. Ohta, Y. Hiei, T. Komari and T. Kumashiro. 1996. High efficiency transformation of maize (*Zea mays* L.) mediated by *Agrobacterium tumefaciens*. Nature Biotechnol. 14: 745–750.

Jackson, M.A., D.J. Anderson and R.G. Birch. 2013. Comparison of *Agrobacterium* and particle bombardment using whole plasmid or minimal cassette for production of high-expressing, low-copy transgenic plants. Transgenic Res. 22: 143–151.

Jain, M., K. Chengalrayan, A. Abouzid and M. Gallo. 2007. Prospecting the utility of a PMI/mannose selection system for the recovery of transgenic sugarcane (*Saccharum* spp. hybrid) plants. Plant Cell Rep. 26: 581–590.

Jefferson, R.A. 1987. Assaying chimeric genes in plants: the GUS gene fusion system. Plant Mol. Biol. Rep. 5: 387–405.

Johnston, S.A., P.Q. Anziano, K. Shark, J.C. Sanford and R.A. Butow. 1988. Mitochondrial transformation in yeast by bombardment with microprojectiles. Science 240: 1538–1541.

Joyce, A., R.B. McQualtert, J.A. Handley, J.L. Dale, R.M. Harding and G.R. Smith. 1998. Transgenic sugarcane resistant to sugarcane mosaic virus. Proc. Aust. Soc. Sugar Cane Technol. 20: 204–210.

Joyce, P., M. Kuwahata, N. Turner and P. Lakshmanan. 2010. Selection system and co-cultivation medium are important determinants of *Agrobacterium*-mediated transformation of sugarcane. Plant cell Rep. 29: 173–183.

Joyce, P., S.H., A. O'Connell, Q. Dinh, L. Shumbe and P. Lakshmanan. 2014. Field performance of transgenic sugarcane produced using *Agrobacterium* and biolistics methods. Plant Biotechnol. J. 12: 411–424.

Jung, J.H., W. Vermerris, M. Gallo, J.R. Fedenko, J.E. Erickson and F. Altpeter. 2012. RNA interference suppression of lignin biosynthesis increases fermentable sugar yields for biofuel production from field-grown sugarcane. Plant Biotechnol. J. 6: 709–716.

Kalunke, R.M., A.M. Kolge, K.H. Babu and D.T. Prasad. 2009. *Agrobacterium*-mediated transformation of sugarcane for borer resistance using *Cry 1Aa3* gene and one-step regeneration of transgenic plants. Sugar Tech. 11: 355–359.

Klee, H.J., R.B. Horsch, M.A. Hinchee, M.B. Hein and N.L. Hoff-mann. 1987. The effects of overproduction of two Agrobacterium *tumefaciens* T-DNA auxin biosynthetic gene products in transgenic petunia plants. Genes Dev. 1: 86–96.

Klein, T.M., E.C. Harper, Z. Svab, J.C. Sanford, M.E. Fromm and P. Maliga. 1988. Stable genetic transformation of intact Nicotiana cells by the particle bombardment process. Proc. Natl. Acad. Sci. USA 85: 8502–8505.

Lam, E., J. Shine, Jr., J. Da Silva, M. Lawton, S. Bonos, M. Calvino, H. Carrer, M.C. Silva-Filho, N. Glynn, Z. Helsel, J. Ma, E. Richard, Jr., G.M. Souza and R. Ming. 2009. Improving sugarcane for biofuel: engineering for an even better feedstock.

Lakshmanan, P., R.J. Geijskes, K. Aitken, C.L.P. Grof, G.D. Bonnet and G.R. Smith. 2005. Invited review: Sugarcane biotechnology: the challenges and opportunities. *In Vitro* Cell Developmental Biology 41: 345–363.

Leibbrandt, N.B. and S.J. Snyman. 2003. Stability of gene expression and agronomic performance of a transgenic herbicide-resistant sugarcane line in South Africa. Crop Science 43: 671–677.

Li, Q., L. Fan, Q. Luo, H. He, J. Zhang, Q. Zeng, Y. Li, W. Zhou, Z. Huang, H. Deng and Y. Qi. 2014. Co-overexpression of AtCBL9, AtCIPK23 and AtAKT1 enhances K$^+$ uptake of sugarcane under low-K$^+$ stress. Plant Omics J. 7: 188–194.

Liu, D., S.V. Oard and J.H. Oard. 2003. High transgene expression levels in sugarcane (*Saccharum officinarum* L.) driven by the rice ubiquitin promoter *RUBQ2*. Plant Science 165: 743–750.

Ma, H., H.H. Albert, R. Paull and P.H. Moore. 2000. Metabolic engineering of invertase activities in different subcellular compartments affects sucrose accumulation in sugarcane cells. Aust. J. Plant Physiol. 27: 1021–1030.

Maliga, P. and R. Bock. 2011. Plastid Biotechnology: Food, Fuel, and Medicine for the 21st Century. Plant Physiol. 155: 1501–1510.

Manickavasagam, M., A. Ganapathi, V.R. Anbazhagan, B. Sudhakar, N. Selvaraj, A. Vasudevan and S. Kasthurirengan. 2004. *Agrobacterium*-mediated genetic transformation and development of herbicide-resistant sugarcane (*Saccharum* species hybrids) using axillary buds. Plant Cell Reports 23: 134–143.

Mayavan, S., S. Kondeti, M. Arun, M. Rajesh, G.K. Dev, G. Sivanandhan, B. Jaganath, M. Manickavasagam, N. Selvaraj and A. Ganapathi. 2013. *Agrobacterium tumefaciens*-mediated in planta seed transformation strategy in sugarcane. Plant Cell Reports 10: 1557–1574.

McCormick, A.J., D.A. Watt and M.D. Cramer. 2009. Supply and demand: sink regulation of sugar accumulation in sugarcane. Journal of Exp. Botany 60: 357–364.

McQualter, R.B., J.L. Dale, R.M. Harding, J.A. McMahon and G.R. Smith. 2004a. Production and evaluation of transgenic sugarcane containing Fiji disease virus (FDV) genome segment S9-derived synthetic resistance gene. Aust. J. Agricult. Res. 55: 139–145.

McQualter, R.B., B. Fong Chong, M. O'Shea, K. Meyer, D.E. van Dyk, P.V. Viitanen and S.M. Brumbley. 2004b. Initial evaluation of sugarcane as a production platform for a p-hydroxybenzoic acid. Plant Biotechnol. J. 2: 1–13.

Molinari, H.B.C., C.J. Marura, E. Daros, M.K.F. de Camposa, J.F.R.P. de Carvalhoa, J.C. Bespalhok Filho, L.F.P. Pereira and L.G.E. Vieira. 2007. Evaluation of the stress-inducible production of proline in transgenic sugarcane (*Saccharum* spp.): osmotic adjustment, chlorophyll fluorescence and oxidative stress. Physiol. Plant 130: 218–229.

Mudge, S.R., K. Osabe, R.E. Casu, G.D. Bonnett, J.M. Manners and R.G. Birch. 2009. Efficient silencing of reporter transgenes coupled to known functional promoters in sugarcane, a highly polyploidy crop species. Planta 229: 549–558.

Murashige, T. and F. Skoog. 1962. A revised medium for rapid growth and bioassays with tobacco tissue cultures. Physiol. Plant 15: 473–497.

Nehra, N.D., R.N. Chibbar, N. Leung, K. Caswell, C. Mallard, L. Stein-hauer, M. Baga and K.K. Kartha. 1994. Self-fertile transgenic wheat plants regenerated from isolated scutellar tissues following microprojectile bombardment with two distinct gene constructs. Plant J. 5: 285–297.

Nickell, L.G. 1964. Tissue and cell culture of sugarcane: Another research tool. Hawaii Pl. Rec. 57: 223–229.

Osabe, K., S.R. Mudge, M.W. Graham and R.G. Birch. 2009. RNAi Mediated Down-Regulation of PDS Gene Expression in Sugarcane (*Saccharum*), a Highly Polyploid Crop. Tropical. Plant Biol. 2: 143–148.

Potrykus, I. 1990. Gene transfer to cereals: an assessment. Nature. Biotechnol. 8: 535–542.

Rae, A.L., M.A. Jackson, C.H. Nguyen and G.D. Bonnett. 2009. Functional specialization of vacuoles in sugarcane leaf and stem. Trop. Plant Biol. 2: 13–22.

Rathus, C., R. Bower and R.G. Birch. 1993. Effects of promoter, intron and enhancer elements on transient gene expression in sugar-cane and carrot protoplasts. Plant Mol. Biol. 23: 613–618.

Rathus, C. and R.G. Birch. 1992. Stable transformation of callus from electroporated sugarcane protoplasts. Plant Sci. 82: 81–89.

Reis, R.R., B.A.D.B. da Cunha, P.K. Martins, M.T.B. Martins, J.C. Alekcevetch, A. Chalfun-Júnior, A.C. Andrade, A.P. Ribeiro, F. Qind, J. Mizoie, K. Yamaguchi-Shinozakie,

K. Nakashimad, J.F.C. Carvalho, C.A.F. de Sousa, A.L. Nepomuceno, A.K. Kobayashia and H.B.C. Molinari. 2014. Induced over-expression of *AtDREB2ACA* improves drought tolerance in sugarcane. Plant Science 221-222: 59–68.

Rogalski, M. and H. Carrer. 2011. Engineering plastid fatty acid biosynthesis to improve food quality and biofuel production in higher plants. Plant Biotechnology Journal 9: 554–564.

Rossouw, D., S. Bosch, J. Kossmann, F.C. Botha and J.-H. Groenewald. 2007. Downregulation of neutral invertase activity in sugarcane cell suspension cultures leads to a reduction in respiration and growth and an increase in sucrose accumulation. Funct. Plant Biol. 34: 490–498.

Sainz, M.B. 2009. Commercial cellulosic ethanol: the role of plant-expressed enzymes. *In Vitro* Cell Dev. Biol.-Plant 45: 314–329.

Sanford, J.C., T.M. Klein, E.D. Wolf and N. Allen. 1987. Delivery of substances into cells and tissues using a particle bombardment process. J. Part Sci. Technol. 5: 27–37.

Sanford, J. 2000. Turning point article the development of the biolistic process. *In Vitro* Cell Dev. Biol.-Plant 36: 306–308.

Souza, G.M., H. Berges, S.S. Bocs, R.D. Casu, A. Berg, J.E. Ferreira, R. Henry, R. Ming, B. Potier, M.A. Van Sluys, M. Vincentz and A.H. Paterson. 2011. The Sugarcane Genome Challenge: Strategies for Sequencing a Highly Complex Genome. Tropical. Plant Biology 4: 145–156.

Svab, Z.P. and P. Maliga. 1993. High-frequency plastid transformation in tobacco by selection for a chimeric *aad*A gene. Proc. Natl. Acad. Sci. 90: 913–917.

Taparia, Y., W.M. Fouad, M. Gallo and F. Altpeter. 2012. Rapid production of transgenic sugarcane with the introduction of simple loci following biolistic transfer of a minimal expression cassette and direct embryogenesis. *In Vitro* Cell Dev. Biol.-Plant 48: 15–22.

Tomkins, J.P., Y. Yu, H. Miller-Smith, D.A. Frisch, S.S. Woo and R. Wing. 1999. A bacterial artificial chromosome library for sugarcane. Theor. Appl. Genet. 99: 419–424.

van der Merwe, M.J., J.H. Groenewald, M. Stitt, J. Kossmann and F.C. Botha. 2010. Down-regulation of pyrophosphate: Fructose 6-phosphate 1-phosphotransferase activity in sugarcane culms enhances sucrose accumulation due to elevated hexose-phosphate levels. Planta 231: 595–608.

van der Vyver, C., T. Conradie, J. Kossmann and J. Lloyd. 2013. *In vitro* selection of transgenic sugarcane callus utilizing a plant gene encoding a mutant form of acetolactate synthase. *In Vitro* Cell Dev. Biol.-Plant 49: 198–206.

Vettore, A.L., F.R. da Silva, E.L. Kemper, G.M. Souza, A.M. da Silva, M.I. Ferro, F. Henrique-Silva, E.A. Giglioti, M.V. Lemos, L.L. Coutinho, M.P. Nobrega, H. Carrer, S.C. Franca, M. Bacci, Junior, M.H. Goldman, S.L. Gomes, L.R. Nunes, L.E. Camargo, W.J. Siqueira, M.A. Van Sluys, O.H. Thiemann, E.E. Kuramae, R.V. Santelli, C.L. Marino, M.L. Targon, J.A. Ferro, H.C. Silveira, D.C. Marini, E.G. Lemos, C.B. Monteiro-Vitorello, J.H.M. Tambor, D.M. Carraro, P.G. Roberto, V.G. Martins, G.H. Goldman, R.C. de Oliveira, D. Truffi, C.A. Colombo, M. Rossi, P.G. de Araujo, S.A. Sculaccio, A. Angella, M.M.A. Lima, V.E. de Rosa, Jr., F. Siviero, V.E. Coscrato, M.A. Machado, L. Grivet, S.M.Z. Di Mauro, F.G. Nobrega, C.F.M. Menck, M.D.V. Braga, G.P. Telles, F.A.A. Cara, G. Pedrosa, J. Meidanis and P. Arruda. 2003. Analysis and functional annotation of an expressed sequence tag collection for tropical crop sugarcane. Genome Res. 13: 2725–2735.

Vickers, J.E., C.P.L. Grof, G.D. Bonnett, P.A. Jackson, D.P. Knight, S.E. Roberts and S.P. Robinson. 2005. Overexpression of Polyphenol Oxidase in Transgenic Sugarcane Results in Darker Juice and Raw Sugar. Crop Science 45(1): 354–362.

von Montagu, M. 2011. It is a long way to GM Agriculture. Annu. Rev. Plant Biol. 62: 1–23.

Weng, Li-Xing., Haihua. Deng., Jin-Ling. Xu, Qi. Li, Lian-Hui. Wang, Zide. Jiang, Hai Bao. Zhang, Qiwei. Li., Lian-Hui. Zhang. 2006. Regeneration of sugarcane elite breeding lines and engineering of stem borer resistance. Pest Management Sciences 62: 178–187.

Weng, Li-Xing., Hai-Hua. Deng, Jin-Ling, Jin-Ling. Xu, Qi. Li, Yu-Qian. Zhang, Zi-De, Jiang, Chen, Qi-Wei, Jian-Wen. Chen, Lian-Hui. Zhang. 2011. Transgenic sugarcane plants expressing high levels of modified cry1Ac provide effective control against stem borers in field trials. Transgenic Res. 20: 759–772.

Wu, L. and R.G. Birch. 2007. Doubled sugar content in sugarcane plant modified to produce a sucrose isomer. Plant Biotechnol. J. 5: 109–117.

Zale, J.M., S. Agarwal, S. Loar and C.M. Steber. 2009. Evidence for stable transformation of wheat by floral dip in *Agrobacterium tumefaciens*. Plant Cell Rep. 28: 903–913.

Zhang., S.Z., B.P. Yang, C.L. Feng, R.K. Chen, J.P. Luo, W.W. Cai and F.H. Liu. 2006. Expression of the *Grifolafrondosa trehalose synthase* gene and improvement of drought-tolerance in sugarcane (*Saccharum officinarum* L.). J Integr. Plant Biol. 48: 453–459.

Zhu, Y.J., H. McCafferty, G. Osterman, S. Lim, R. Agbayani, A. Lehrer, S. Schenck and E. Komor. 2011. Genetic transformation with untranslatable coat protein gene of sugarcane yellow leaf virus reduces virus titers in sugarcane. Transgenic Res. 20: 503–512.

CHAPTER 6

Sugarcane Ethanol in Brazil: Socio-Economic Issues

Márcia Azanha Ferraz Dias de Moraes* and Fabíola Cristina Ribeiro de Oliveira

ABSTRACT

In a context of increasing production and use of sugarcane ethanol in Brazil and in the world, it is important to underscore the socioeconomic aspects involved in its production. Although advances in social conditions have been implemented over the past decades, the image of the sugarcane sector to the Brazilian society, as well as for foreigners, only began to improve in recent years. This change is largely due to the increase of public awareness and campaigns for the clarification of producers' commitment to the improvement of labor conditions. In this chapter we examine the main legislation governing the Brazilian labor market; the evolution of the socioeconomic indicators of the productive sectors of sugarcane, sugar and ethanol; the

Departamento de Economia Administração e Sociologia, Universidade de São Paulo, ESALQ, Av. Pádua Dias, 11, 13418900 - Piracicaba, SP, Brasil.
* Corresponding author

role of the academic sector in carrying out studies that allow a better understanding of the current working conditions. From the mid-2000's onwards, pressures imposed by the society and consumers, openness to foreign markets, as well as a stronger enforcement of existing standards and legislation led to the improvement in the labor market as judged by multiple criteria.

The importance of analyzing socio-economic aspects of the production of sugarcane and its products is justified by the fact that the soil and climatic conditions in Brazil are also found in many underdeveloped and developing countries with a large, poorly-educated population that are excluded from the labor market. Thus, for appropriate public policies that can help reduce greenhouse gases emissions while also to add or retain thousands of workers in the labor market, the analysis of the Brazilian experience with sugarcane-based sectors can provide invaluable lessons for police makers and producers alike.

Introduction

The 35 years of Brazilian experience with production and use of ethanol fuel from sugarcane has been widely discussed in literature, and mostly, its positive contribution to reducing CO_2 emissions compared to gasoline. The chief concern of major centers and agencies, as well as the current discussions of modern societies on alternative sources of energy, revolves round the need to find renewable fuels that are more environment friendly, given the negative impacts caused by the production and use of fossil fuels.

Few works deal with social issues involved in the sugarcane sector, namely, job creation (agricultural and industrial sectors), working conditions and evolution in socio-economic indicators. Moreover, part of the literature on social issues regarding sugarcane production in Brazil is quite critical, often placing emphasis on the long-existing inappropriate working conditions throughout the history of this industry. Other authors, on the other hand, working with officially published data, have shown significant improvements in working conditions since 2000, and indicated that some statistics (formal employment, schooling, wages) are even better in this sector as compared to other agricultural activities.

Society's perception of ethanol production from sugarcane is influenced by available information, both in scientific papers and especially in the media. Although advances in social conditions have been implemented over the last few decades, the image of the sector to the society began to improve only in recent years due to the increase of information available (either by the larger number of scientific papers published on the issue, reports released in the press and campaigns for the clarification of producers' shares of commitment to the sector).

Thus, aiming to contribute to this debate, the objectives of this chapter are: (i) to present aspects of the institutional environment (the main legislation governing the Brazilian labor market) in order to highlight the existing legal apparatus in the sugarcane sector; (ii) to analyze the evolution of the socio-economic indicators of the productive sectors of sugarcane, sugar and ethanol (carried out based on data from two official databases of the Brazilian Government: PNAD and RAIS[1]); (iii) by a literature review, to capture the role of the academic sector in carrying out studies that allow a better understanding of the current working conditions of workers in the sugarcane sector; (iv) to analyze the initiatives of enterprises regarding improvements to the sector with the objective to enlighten the public.

The importance of analyzing aspects of the production of sugarcane, sugar and ethanol is justified by the fact that the soil and climatic conditions in Brazil, which make sugarcane ethanol production in this country extremely competitive, are also found in many underdeveloped and developing countries with a large, poorly-educated population that are excluded from the labor market. In a given scenario of appropriate public policies that not only aim to reduce greenhouse gases emissions, but to also include thousands of workers to the labor market, the Brazilian experience could be replicated, thus generating jobs and income, while contributing to mitigate CO_2 emissions in the world.

RESULTS AND DISCUSSION

Labor Market Regulations

The main regulatory laws for the labor market are imposed by the government and conventions defined by the workers' unions and the industry syndicate, which are juridical regulations provisioned in the Labor Law Code.

The regulatory laws for the labor market in Brazil are: (i) The Federal Constitution; (ii) Consolidation of Labor Laws (CLT), (iii) Rural Workers' Law (5889/73; and Law No. 10.192/2001) that establishes the wage policy.[1]

---

[1] This law forbids to specify or to fix an adjustment or automatic wage correction clause attached to price indexes in the negotiations between the Industry Syndicate and the Workers' Union and the Collective Agreements. The law also prescribes that any concession of salary raise due to productivity increase shall be supported by objective indicators.

In addition, there are annual negotiations between the parties (Collective Agreements) held in compliance with the above mentioned laws.

The Brazilian labor legislation is rather strict, however, it allows flexibility through the collective agreements between the parties in two aspects: wage and working hours, which must be negotiated in compliance with the labor law, through the Collective Agreements.

There are two types of Collective Agreements: the Collective Labor Conventions, involving negotiations between the Industry Syndicate and the Workers' Union, and the Collective Agreements, when enterprises negotiate directly with the Workers' Union. Should there be divergences in the negotiations settled by both, the prevailing agreement must favor the workers.

The Federal Constitution (FC/Magna Carta) of Brazil proclaims to be social rights: education, health, work, leisure, safety, social security, maternity and childhood protection, and assistance to the helpless (FC, article 6). It is the country's major law to which all others are subjected. The organization of workers' unions is covered in article 5 of the Constitution.

With regard to the system of the workers' unions, it is composed of three levels: municipal, state and national. At the municipal level, the category is represented by the workers' unions (one union can represent several neighboring municipalities). The unions are usually affiliated to a state organization, called Federation. The Federations, in turn, are usually affiliated nationwide to the Confederation.

Moraes (2011) analyzed the determinants of employees' income of the sugar industry, especially in terms of the union's activity.[2] The author reported on the existence and activity of unions, as well as their influence on wage formation.

Thus, there is an extensive legal and regulatory apparatus governing the Brazilian labor market, covering all sectors of the economic activity, including workers in the sugarcane, sugar and ethanol sectors.

---

[2] The author estimated equations for the income of workers in the sugarcane, sugar and ethanol sectors and collected data on the workers' union representation by a questionnaire applied to workers. In the estimated earning regressions, the coefficients of the following variables were significant and expected: (i) gender, (ii) region, (iii) schooling, (iv) threshold effect of education, (v) membership of workers' union.

Evolution of Socio-Economic Indicators

Data from the Annual Report of Social Information (RAIS), published by the Ministry of Labor and Employment; and from the National Household Sample Survey (PNAD), conducted by the Brazilian Institute of Geography and Statistics (IBGE),[3] were used to analyze the evolution of the number of people employed, labor formalization, schooling, and wages in sugarcane crops and in ethanol and sugar production.

The evolution of the number of employees formally registered,[4] in the three sectors: sugarcane, sugar and ethanol, considering the RAIS database, is presented in Table 1. The data are presented separately for the two producing regions: North-Northeastern (NNE) and Central-Southern (CS) regions, whose strategies and productivity levels differ greatly. Sugar production is the main activity of the North-Northeastern region, while

Table 1. BRAZIL: Number of formal workers*by producing regions and sector, between 2000 and 2010.

	Region	2000	2010
Sugarcane	NNE	81,191	94,153
	CS	275,795	325,907
	Total for Brazil	356,986	420,060
Sugar	NNE	143,303	242,205
	CS	74,421	306,473
	Total for Brazil	217,724	548,678
Ethanol	NNE	25,730	47,759
	CS	42,408	165,349
	Total for Brazil	68,138	213,108
Total for Brazil (3 sectors)	642,848		1,181,846

Source: Table built based on RAIS data (2000 and 2010)

---

[3] RAIS is an annual census on the formal labor market based on information provided by enterprises. Data are available considering the sector of the activity at municipal level. PNAD is an annual survey of socioeconomic information on formal and informal workers, obtained through questionnaires applied to a sample number of households; data are available considering the sector of the activity at state level.

[4] All employees have labor rights prescribed in the legislation.

ethanol production used to be the main activity of the Central-Southern region, however, sugar production has been growing significantly in recent years.

It should be noted that, for Brazil as a whole, considering the three sectors (sugarcane, sugar and ethanol) there has been a significant increase (83.8%) in the number of formal workers between 2000 and 2010 (from 642,848 workers in 2000 to 1,181,846 in 2010), which can be explained by the significant production growth in the period.[5] When considering the total number of formal workers, in the three sectors, we observe that about 67.5% was located in the Central-Southern region of the Country in 2010. It is seen that the increase in the number of formal workers in ethanol mills (212.8%) and in sugar processing plants (311.8%) was higher than that of rural workers (17.7%) involved in sugarcane production, probably because of the shift to the mechanization process of sugarcane harvesting. It should also be noted that, in 2000, about 55% of workers were working in sugarcane crops and this figure dropped to 35.5% in 2010.

The proportion of workers hired informally is usually higher in the agriculture sector.[6] Therefore, PNAD database, which presents information on all workers (formal and informal ones), is suitable when analyzing the socio-economic indicators of agriculture sectors. According to the PNAD data, in 2009, there were 2,642,261 agricultural employees in Brazil, and sugarcane production accounts for 21% of this total. In addition, the sector employs many poorly educated people, who would be out of the job market otherwise. Table 2 shows some indicators for workers in sugarcane crops.

Table 2. Indicators for workers in sugarcane crops – 2009.

	Number of Workers	Average Age (years)	Monthly Average Wage* (US$)	Male Share	Average Schooling (years)
Brazil	542,588	34.9	204.35	92%	4.5
North-Northeast	246,806	33.2	147.92	95%	3.4
Central-South	295,782	36.2	251.51	87%	5.5
São Paulo state	175,021	36.5	269.70	85%	6.0

Source: Table built based on PNAD data, 2009
* US$ exchange rate of Sept. 2009

---

[5] Sugarcane production moved up from 326.12 million tons to 722.5 million; ethanol production has risen from 4 to 28 billion liters, and sugar production from 7.8 to 26.8 million tons.
[6] In the sugar and ethanol sectors, roughly 100% of workers are formal employees.

The PNAD data, for the year 2009, show that the sugarcane sector employs 542,588 workers.[7] Moreover, the data show that workers in sugarcane crops are shared in the two producing regions at approximately the same percentage. Although it is the main producing state, accounting for nearly 60% of the production, São Paulo state represents only 32.2% of all jobs in the sugarcane sector, due to the implementation of the mechanized systems for harvesting and planting sugarcane in that state, which demands less workforce.

No significant differences were observed in terms of average age of workers in the two producing regions. Regarding gender, most workers are men in the two regions, while the percentage of women is higher in the Central-Southern region (13%), which can be attributed to the mechanization of agricultural activities in this region.

Wage averages are higher in São Paulo state. In 2009, the average wage paid to workers in São Paulo was 32% higher than that in the other producing regions of the country.

The number of workers employed for sugarcane crops has dropped since 1981[8] (Fig. 1). The data show that in 1981, the sector employed more than 625 thousand employees. In 1985 the number of workers was the highest, and has gradually been decreasing since then. In 2009, there were about

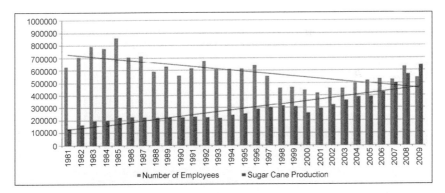

Figure 1. Evolution of the number of employees and sugarcane production. Source: Built based on PNAD (several years) and MAPA data.

---

[7] Excluding self-employed workers, producers of sugarcane for their own consumption and non-remunerated producers.

[8] Brazilian National Ethanol Program—Proalcool—was launched in 1975, to stimulate the production of *anhydrous ethanol* (to be used mixture to gasoline) and *hydrous ethanol* (to be used in automobiles powered by hydrous ethanol and on flex fuel cars). PNAD data started to be collected in 1981.

542.5 thousand employees in the sector, which means a 13.3% reduction. In the same period (1981–2009), the sugarcane production increased from 154 million tons to 640 million tons, a roughly 315.6% growth.

These opposing trends are attributed to a rise in labor productivity, and mainly due to the implementation of the mechanized sugarcane harvesting process (Fig. 2) in Sao Paulo state and in the new sugarcane producing regions of the central-western region.

In addition to the legislation that bans sugarcane burning in the state of Sao Paulo, UNICA (a Sugarcane Industry Association), which represents most of the sugar and ethanol producers in São Paulo state, and the São Paulo state government cosigned an Environmental Protocol that established the deadlines for the elimination of sugarcane burning, and therefore, increased the mechanization of agricultural activities in the state. Thus, the protocol prescribes 2014 as the deadline to terminate sugarcane burning in the mechanizable areas (2021 by law), and 2017 in non-mechanizable areas (2031 by law). The termination of sugarcane burning reduces the demand for manual labor and changes workers' profile, besides the implication of a restructuring of agricultural labor.

Data show that there is a trend for the formalization of the activity in the sugarcane sector. The formalization of the activity indicates that the working conditions are good and in compliance with the current legislation.

Figure 2. Mechanical Sugarcane Harvesting. Source: Courtesy of União da Indústria de Cana-de-Açúcar - TadeuFessel.

The number of workers formally employed in the agricultural sector in Brazil is relatively small. In 2008, it accounted for nearly 36.7% of the workforce. However, the situation is much better for the sugarcane sector where roughly 80.7% of the employees were formally employed (formal labor contract) in 2009. The formal employees are entitled to all labor rights prescribed in the legislation, such as the right to receive unemployment insurance; annual paid-vacation; bonus salary (aka 13th salary in Brazil), among others.

The rate of formal and informal workers in the two producing regions diverges. In the North-Northeast, almost 72% of workers are formal (higher than the agricultural sector as a whole). The largest number of formal workers is in São Paulo state—Brazil's major sugarcane producer—formally employing 175,021 workers by contracts (accounting for 94.8% of the total workers of the sugarcane sector in this state).

The evolution observed in the distribution of the number of employees on the basis age bracket in the 1981–2009 period is shown in Table 3. The data show that the participation of employees at the age of 15 or under was 15.3% (approximately 95.5 thousand employees) in 1981 and that it dropped to 0.3% in 2009 (totaling roughly 1,641 employees). This indicates a drastic reduction of child labor for sugarcane production in Brazil.

In conclusion, we can see that the sugarcane sector successfully reduced child labor in the sector and increased the proportion of formal adult workers. These results show a combination of efforts, such as governmental

Table 3. Sugarcane: evolution of the number of employees by age bracket.

Age bracket	1981		2009	
	Number of Employees	Total %	Number of Employees	Total %
10–15	95,576	15.3	1,642	0.3
15–20	128,578	20.6	44,492	8.2
20–30	134,033	21.4	187,374	34.5
30–40	106,516	17	141,859	26.1
40–50	84,041	13.5	107,130	19.8
50–60	51,886	8.3	46,806	8.6
> 60	24,386	3.9	13,285	2.5
Total	625,016	100	542,588	100

Source: Table built based data provided by the PNAD (1981 and 2009)

social programs (the Family and School Grants programs); the commitment of sugarcane producers to improvements; better law observance and requirements by the market itself.

Regarding schooling, data show poor indicators. The average schooling level of workers in the agricultural sector in Brazil as a whole was 4 years in 2009. In the sugarcane sector, the workers had about 4.5 years of study on average. The state of São Paulo presented better indicators (5.1 years of school on average), although still very low.

Despite the positive evolution observed in the schooling level of workers in the sugarcane industry in Brazil between 1981 and 2009 (average schooling increased from 2.2 to 4.5 years), it should be highlighted that it is still very low: 42.8% of the workers for sugarcane crops in Brazil in 2009 had at the most 4 years of schooling, of which about 21.4%, i.e., 116 thousand workers, are illiterate.

The Role of the Academic Sector and the Dissemination of Information

In several moments along the history of sugarcane production in Brazil, the producers of sugarcane, sugar and ethanol have been in the spotlight due to complaints concerning pollution of air and rivers; exploitation of the workforce; allegations of an expansion into forested areas or into areas previously used for food crops. These criticisms feed the public's imagination and overshadow the potential of the sector as a generator of income and employment (Cunha and Silva 2011).

As demonstrated earlier, this sector employs directly and formally more than 1 million people, including workers with low education levels. However, the working conditions of employees involved in sugarcane production are the aspects that gained prominence recurrently in the media such as newspapers and television. This scenario began to change from the mid-2000s, probably due to improvements in working conditions and to the increase of information about the sector to the public.

The performance-based income, predominant in the manual harvesting of sugarcane, according to some authors, is directly associated with work-related accidents and deaths that occur in sugarcane crops (Balsadi 2007; Alves 2006). More incisively, Alves (2006) states that the production processes and working conditions existing in the sugarcane crop sector, the constant demand for increasing productivity combined with the performance-based income, push workers to the limit of efforts employed at work. The increased energy demand and effort to cut more cane can cause the death of workers or the loss of working capacity, according to the author.

On this issue, two aspects should be highlighted: (1) the performance-based income has been used in activities that still depend on the speed and skill of the worker, such as harvesting (Fig. 3). Therefore, other sectors such as orange, lemon, tangerine, coffee and cotton, also adopt this remuneration system (Baptistella et al. 1994; Oliveira 2009); (2) to date, there are no scientific studies demonstrating the causal link between the performance-based income and workers' deaths in the sugarcane industry in Brazil (Moraes 2007).

Furthermore, based on data from RAIS, Moraes and Ferro (2011) showed that the numbers of deaths and retirements by work-related accident in sugarcane production in 2009, besides being small as compared to the total number of workers (0.003%) are at the same levels as those in the agricultural sector as a whole.

Still, since 2006, the Ministry of Labor of the 15th Region of São Paulo state has tried to eliminate the performance-based remuneration system for workers of the sugarcane production sector, attributing to this system the cause of workers' deaths. It is a controversial intervention, given that the decision for the termination of the performance-based system is not a consensus among the workers' unions. A section of workers is in favor of this remuneration system. UNICA also shows an opposing position

Figure 3. Manual Sugarcane Harvesting. Source: Courtesy: União da Indústria de Cana-de-Açúcar (UNICA) - Andreas Niels.

concerning the termination of performance-based remuneration system, but the association highlights its commitment to reinforcing the compliance with existing labor laws by the sugarcane producers to ensure the rightful remuneration for workers, as stipulated in the collective labor agreements (Moraes 2007). The importance of this issue has reduced with the adoption of mechanized process of sugarcane harvesting (that accounts for 98.3% of the 2013/2014 crop year in the Center-South region of Brazil), because the performance based remuneration is not adopted in this case.

Nevertheless, there are still cases of noncompliance with labor laws regarding the sugarcane cutters, yet representing a very small percentage of the total number of workers employed in the industry. Even so, these situations receive a great deal of attention from the media and end up contributing to the perception that the working conditions for the employees of the sector are still poor.

Oliveira (2009) shows that the sugarcane industry has the highest number of formal workers, compared to other agricultural activities such as rice, coffee, cassava, corn, and even soy. In 2007, 80.9% of employees in the sugarcane sector were registered, while in the cassava and corn sectors, this figure was only 2.5% and 9.1%, respectively. It is a fact that the proportion of registered employees in the sugarcane sector is increasing and this may be related to the intensification of inspection by governmental institutions and to the organization of the workers' union movement. Furthermore, the vision of entrepreneurs in the sugarcane industry has changed, and it is no longer the same as that of the past feudal lords who enslaved workforce in order to have higher production and profits. The social and environmental issues, along with a more rigid position on the part of consumers, led to the adoption of social and environmental programs by the producers, which have become embedded in the management procedures of sugarcane mills and gained importance within the administrative processes (Cunha and Silva 2011).

It is important to recognize that the sugarcane industry employs a large number of people with a low level of education. Therefore, without this inclusion in the job market, these workers would probably have difficulties to be absorbed by other productive sectors.

The perception that the population in general holds about the sugarcane industry, usually broadcast by the media in Brazil, is different from the reality. Bragato (2008) investigated the perception of city residents directly benefited by the social measures developed by some sugarcane mills in São Paulo state. The author found that issues related to job opportunities, income creation, and regional development express the image about sugarcane mills created in these communities better. Assato et al. (2011) carried out interviews with several social actors, either from the public or from the private sectors, to indentify the impacts arising from the installation of sugarcane mills in the municipalities of Alvorada do Sul and Rio Brilhante, both in Mato Grosso do Sul state—Brazil. The authors found that the perception of all respondents is that there have been improvements in terms of employment, income, and development of other commercial and productive activities in the municipalities. This is in addition to an expansion of the educational system, with the implementation of technical schools in partnerships with the industrial sector and governmental agencies.

It is worth noting that much of the past criticism made to the sugarcane sector by the media and researchers has contributed to the advancement and improvement of working conditions in the sector. In 2008, the "Mesa de Diálogo para Aperfeiçoar as Condições de Trabalho na Cana-de-Açúcar" (Round Table for the Improvement of Working Conditions in the Sugarcane Industry) was installed and coordinated by the General Secretariat of the Presidency. This round table was comprised of workers' representatives (through the National Confederation of Agricultural Workers (Contag), the Federation of Rural Workers of São Paulo State (Feraesp)) and entrepreneurs of the sector affiliated to UNICA, and by federal government representatives. Since its creation, several meetings have been held and in June 2009, in the presence of the former Brazilian President Luiz Inacio Lula da Silva, ministers, unionists and businessmen of the sector, signed the document entitled "National Commitment to Improve Working Conditions in the Sugarcane Sector". This obtained voluntary compliance by almost all sugarcane mills in operation in the country. It is noteworthy that the following points were agreed upon: guarantee of hiring workers directly (not through third parties), and improved health and safety conditions at work.

Since its signing, regular meetings have been held to make improvements to the protocol, and there is a follow-up to check the compliance with the protocol terms by the Federal Government, which certainly contributes positively to the improvement of working conditions in the sugarcane crop sector. The document also states the participation of sugar and alcohol companies in training and qualification of workforce, which is important in a scenario of increasingly mechanized process of sugarcane harvesting.

Moraes (2007) believes that the expansion of the sector in recent years in Brazil can open up good prospects and job opportunities for many workers. However, the new jobs will require skilled workers, such as tractor drivers, drivers, mechanics, operators of harvesters and agricultural machinery as a whole, agricultural technicians and agronomists, among others, given the expansion of mechanized harvesting activities and sugarcane cropping.

There will, however be, a reduction of the demand for employees with low education, the author predicts.

In addition, Fredo et al. (2008) claim that it is still difficult to predict how the workforce involved with manual cutting of sugarcane will be relocated, whether within the sugar/ethanol sector, to other agricultural activities or even other economic sectors. The authors also believe that a portion of the workers will not be relocated within the sector or even to other sectors, given their low level of qualification.

This reality points to the need of public and private actions to provide the completion of formal education and (re)qualification of workers from the sugarcane industry in times of technological advances adopted in Brazilian agriculture.

The Role of the Academic Sector in this Debate

As it is noted, the sugarcane industry in Brazil has a great potential to generate wealth, playing a significant social and economic role in the country's economy. Despite that, the negative information broadcasted by segments of the media industry, particularly on the working conditions of sugarcane cutters, do not always reflect the reality of the sugarcane industry as a whole, and many times local issues are often portrayed as a national reality.

Given the need to further discuss aspects of working conditions (child labor, formal employment, performance-based remuneration, outsourcing, employees' migration, etc.); aspects regarding institutional changes (the banning of burning of sugarcane crops in the State of Sao Paulo and impacts on the labor market) and the effective compliance with laws for the agricultural labor market, several studies have been carried out by researchers from several institutions, for example, those developed by GEMT[9]—Grupo de Extensãoem Mercado de TrabalhoAgrícola (Extension Group in the Agricultural Labor Market)—the University of Sao Paulo - USP - Brazil.

Among the studies conducted by researchers of this group, Moraes and Ferro (2011) analyzed data on death causes and retirement of formal workers in the sugarcane crops and agricultural activities in Brazil as a whole and São Paulo state. The authors showed that the number of workers' deaths in the sector is small compared to the entire workforce and averages the Brazilian agricultural activities as a whole.

---

[9] http:www.esalq.usp.br/gemt

With regard to the retirement, the study indicates that the workers of the sugarcane crops do not retire early (as suggested by some qualitative work). While 0.005% of workers in the sugarcane sector retired after work-related accidents in 2005, in agricultural activities as a whole the average was 0.016% of the total.

Another study conducted by Moraes et al. (2009), which investigated the dynamics of migration of sugarcane workers to the State of São Paulo, showed the importance of the manual cutting activity of sugarcane to migrants from poorer states in the country. The municipalities of Leme (SP) and Pedra Branca (CE) were chosen to be analyzed and characterized as the destiny city and city of origin of workers, respectively. The existence of a single enterprise in the textile sector that employs a small number of workers, along with subsistence agriculture, does not generate jobs and income for the young population of the municipality of Pedra Branca, from where, every year, workers in search of better opportunities migrate to the municipality of Leme to work in the cutting activity of sugarcane in an attempt to include themselves in the labor market.

The work carried out by Hoffmann and Oliveira (2008) showed that the average income of people employed in the sugarcane sector had an actual growth over the period 2002–2006. The growth was 32.4%, while the minimum wage grew 30.9%, indicating that the minimum wage regulates the behavior of remuneration bases in the labor market. The average earnings of employees in the production sector of sugar and alcohol, on the other hand, are higher than the average income of workers in the sugarcane crops; however, from 2002 to 2006 the actual gains of income in these industrial activities were much lower than those verified in the sugarcane crop. Comparing the features and remuneration of workers in sugarcane crops in Brazil with those of other crops (rice, banana, citrus, cassava, soy, corn and grapes), the authors found that the highest average income is observed in soy crops, where most employees are tractor drivers. Income of workers in the sugarcane crops comes in second.

Oliveira (2009) also contributed to showing the evolution of some indicators for employment and remuneration in the sugarcane crops and other agricultural activities in the period between 1992 and 2007. The results show that the increase of modernized agricultural activities is leading to major changes in techniques and social relationships of production. Besides reducing the number of employees, there are qualitative changes in labor force, with indications of a growing demand for employees with higher technical qualifications. It is undeniable the increasing trend of jobs with more stable employment relationships. The formalization of work activities rose in the sugarcane crops and in other crops as well; however, the coefficient in sugarcane crop is higher.

Moraes et al. (2011) highlighted, in the study on socio-economic indicators of the sectors for production of sugarcane, ethanol, oil extraction and oil by products production, the importance of the sugarcane industry in terms of job creation, income and regional development. The authors verified that sugarcane production is located in the countryside of almost all the Brazilian states, spread across many municipalities. Moreover, as the extraction and production of oil by products are limited to a few coastal cities, the sugarcane industry gains visibility for its potential to generate a dynamic regional development.

Other authors have also noted improvements in indicators, such as education, employment (quality and quantity), as well as wages of the mentioned sectors (Balsadi and Borin 2006; Moraes 2007; Oliveira 2009; Moraes 2011; Walter and Machado 2014; Neves and Castro 2013).

Some studies have demonstrated the positive socio-economic impacts of sugarcane ethanol industry in Brazil. Chagas et al. (2011) analyzed the effects of the increasing sugarcane production on municipal revenues of São Paulo state. They showed that the value of agricultural production of sugarcane is higher per hectare than for most crops, thus accruing a greater value of agricultural income to the municipality in terms of tax income. Satolo and Bacchi (2013) assessed the effects of the sugarcane sector expansion over municipal per capita GDP, noting that the GDP for one municipality and that of its satellite neighbours grew from 24% in 2000 to 55% in 2010.

The Private Sector Initiatives

In order to disseminate information about the sugarcane industry, the "Projeto Agora" was launched in 2009 and was established as the greatest initiative for institutional communication of the Brazilian agribusiness. There are 19 associations and companies in the sugarcane industry united in knowledge generation, dissemination of positive social and environmental impacts and, crucially, the provision or extension of insights into the awareness of public opinion regarding issues and sustainability of the sugarcane industry (www.projetoagora.com.br).

In less than two years of existence, the "Projeto Agora" has been recognized as one of the most effective efforts for the dissemination of agri-business. Its activities are directed to the public of most interest: policy makers, consumers, the public, high school students and journalists.

Conclusion

In the context of increasing production and use of sugarcane ethanol in Brazil and the world, it is important to underscore the socio-economic aspects involved in its production. In addition to the solid institutional apparatus that regulates the labor market in Brazil, it is interesting to note the large number of jobs created in the three sectors analyzed: sugarcane crops, sugar and ethanol production, which surpass one million formal jobs.

It is observed that since the creation of the "Proálcool" program in Brazil in 1975, there have been not only an improvement of indicators for the sugarcane sector (working conditions, formal employment, drastic reduction of child labor, investment in training and qualification, social and environmental certification programs adopted by companies, commitment to improving working conditions signed between companies and the Federal Government, etc.), but also in terms of the information available on the subject.

In this context, we verified the importance of gathering and disseminating to the mainstream media (newspapers, magazines, television, radio) data on job creation, positive aspects of the sugarcane industry such as income and wealth generators in thousands of municipalities throughout Brazil, the positive environmental effects of the ethanol production and usage, among others, which was accomplished primarily through the actions of the "Projeto Agora".

Regarding the importance of academic research to the debate about the social aspects of the production of sugarcane, sugar and ethanol, it is noteworthy that initially, most studies were based on criticism to the sector, reflecting the working conditions until the early 2000s. Although they did not reflect the reality of most companies, these studies were of great importance, as they revealed a situation that was not socially sustainable, therefore, they contributed greatly to the necessary changes.

From the mid-2000s onwards, pressures imposed by the society and consumers, openness to foreign markets, as well as a stronger enforcement of existing standards and legislation led to the improvement of working conditions, which were also reflected in the subsequent academic works.

Recent studies have adopted other methods, examining additional important aspects of the labor market such as the impact of institutional changes (notably the effect of banning sugarcane burning in the State of Sao Paulo on employment), analysis of the development generated by the installation of sugarcane mills in the municipalities, comparisons with socio-economic indicators of the oil production sector, analysis of international criteria for social and environmental sustainability, among others.

Undoubtedly, the spread of reliable information contributes to the improvement of working conditions and to a better understanding of the sugarcane industry in Brazil.

References

Alves, F. 2006. Porque morrem os cortadores de cana? Saúde e Sociedade, São Paulo 15(3): 90–98, set./dez.

Assato, M.M., M.A.F.D. de Moraes and F.C.R. de Oliveira. 2011. Impactos sócio-econômicos da expansão do setor bioenergético no Estado do Mato Grosso do Sul: o caso dos municípios de Nova Alvorada do Sul e Rio Brilhante. *In*: Congresso Brasileiro de Economia E Sociologia Rural, 49., 2011, Belho Horizonte. Anais...Brasília: SOBER, 19 p.

Balsadi, O.V. 2007. O mercado de trabalho assalariado na cultura da cana-de-açúcar. DossieEthanol. Revista Eletrônica da SBPC, n. 86. Disponível em:<http://www.comciencia.br/ comciencia>. Acesso em: 07 dez. 2011.

Balsadi, O.V. and M.R. Borin. 2006. OcupaçõesAgrícolas e não-agrícolas no rural paulista-análise no período 1990–2002, São Paulo emPerspectiva, vol. 20 no 4, São Paulo, Brazil.

Baptistella, C. da S.L., M.C.M. Vicente, V.L.F. dos S. Francisco and F.A. Pino. 1994. O trabalho volante na agricultura paulista e sua estacionalidade, 1985-931. Agricultura em São Paulo, São Paulo 41(3): 61–83.

Bragato, I.R. 2008. Percepções dos agentes sociais sobre as práticas de responsabilidade social corporativa em usinas de açúcar e álcool. 141 p. Dissertação (Mestrado Profissional em Administração)—Universidade Metodista de Piracicaba, Piracicaba.

Chagas, A., R. Toneto, Jr. and C. Azzoni. 2011. The expansion of sugarcane cultivation and its impact on municipal revenues: an application of dynamics spatial panels to municipalities in the state of Sao Paulo, Brazil. pp. 137–150. *In*: Edmund Amann, Werner Baer and Don Coes (eds.). Energy, Bio Fuels and Development: Comparing Brazil and the United States. Routledge. Taylor and Francis Group.

Cunha, K.M.R. da and R.D. da Silva. 2011. Bioenergia na mídia: como tornar vantajosa a visibilidade mediada. Bioenergia em revista: diálogos, Piracicaba 1(1): 65–73, jan./jun. 2001. Disponível em: <http://www.fatecpiracicaba.edu.br>. Acesso em: 18 dez.

Fredo, C.E., M.N. Otani, C. da S.L. Baptistella and M.C.M. Vicente. 2008. Recorde na geração de empregos formais no setor agropecuário paulista em 2006. Análise e Indicadores do Agronegócio, São Paulo 3(3): 1–5.

Hoffmann, R. and F.C.R. de Oliveira. 2008. Remuneração e características das pessoas ocupadas na agroindústria canavieira no Brasil, de 2002 a 2006. *In*: Congresso Brasileiro de Economia E Sociologia Rural, 46., 2008, Rio Branco. Anais... Brasília: SOBER, 19 p.

Moraes, M.A.F.D. 2011. Socio-economic Indicators and Determinants of the Income of Workers in Sugar Cane Plantations and in the Sugar and Ethanol Industries in the North, North-East and Centre-South Regions of Brazil. *In*: Edmund Amann, Werner Baer and Don Coes (Org.). Energy, Bio Fuels and Development: Comparing Brazil and the United States: Routledge. Taylor and Francis Group, 334 p.

Moraes, M.A.F.D. de. 2007. O mercado de trabalho da agroindústria canavieira: desafios e oportunidades. Economia Aplicada, São Paulo 11(4): 605–619, out./dez.

Moraes, M.A.F.D. de, C.C. da Costa, J.J.M. Guilhoto, L.G.A. de Souza and F.C.R. de Oliveira. 2011. Social externalities of fuels. pp. 44–75. *In*: E.L.L. de Souza and I. de C. Macedo. (eds.). Ethanol and bioelectricity: sugarcane in the future of the energy matrix. São Paulo: UNICA. cap. 2, Disponível em<http://www.unica.com.br/multimedia/publica.

Moraes, M.A.F.D. de and A.R. Ferro. 2011. Indicators of Mortality and Retirement. Piracicaba, Jul. 2008. 26 p. (Research Report). Available at: <http://www.esalq.usp.br/gemt>. Acessoem 21 nov. 2011.

Moraes, M.A.F.D. de, M.G. de Figueiredo and F.C.R. de Oliveira. 2009. Migração de trabalhadores na lavoura canavieira paulista: uma investigação dos impactos sócio-econômicos nas cidades de Pedra Branca, Estado do Ceará, e de Leme, Estado de São Paulo. Revista de Economia Agrícola, São Paulo 56(2): 21–35, jul./dez.

Neves, M.F. and R.A.O. Castro. 2013. Indústria da Cana: Vetor de Desenvolvimento. Agroanalysis. pgs. 14–15. Julho.

Oliveira, F.C.R. 2009. Ocupação, emprego e remuneração na cana-de-açúcar e em outras atividades agropecuárias no Brasil, de 1992 a 2007. 167 p. Dissertação (Mestrado em Economia Aplicada)—Escola Superior de Agricultura Luiz de Queiroz-, Universidade de São Paulo, Piracicaba.

Pesquisa Nacional Por Amostra DE Domicílios. PNAD. CD-ROM. Rio de Janeiro, RJ. Several years.

Ministério da Agricultura. Pecuária e Abastecimento (MAPA -Ministry of Agriculture, Livestock and Supply). Availabl eat:< http:www.agricultura.gov.br>.

Relação Anual de Informações Sociais (RAIS). Available at: http://www.rais.gov.br/.

Satolo, L.F. and M.R.P. Bacchi. 2013. Impacts of the Recent Expansion of the Sugarcane Sector on Municipal per Capita Income in São Paulo State. Hindawi Publishing Corporation. ISRN Economics Volume 2013, Article ID 828169, 14 pages http://dx.doi.org/10.1155/2013/828169.

União da Indústria de Cana-de-Açúcar– Available at http://unica.com.br/documentos/.

Walter, A., P.G. Machado. 2014. Socio-Economic Impacts od bioethanol from sugarcane in Brazil. pp 193–213. *In*: D. Rutz and R. Janssen (eds.). Socio-Economic Impacts of Bioenergy Production. ISBN: 978-3-319-03828-5 (Print).

Index

About the Volume Editors

Eric Lam is a Distinguished Professor in the Department of Plant Biology and Pathology. His research interests include the study of mechanisms that control programmed cell death and stress tolerance in plants. He is author on over 150 publications in journals including *Science* and *Nature* and has been awarded 5 patents relating to biotechnology methods. He is a recipient of the Alexander von Humboldt award and a Fulbright-Brazil Fellow.

Helaine Carrer is a Full Professor in the Department of Biological Sciences at Agriculture College "Luiz de Queiroz", University of São Paulo. Her research includes Genetic Transformation and Functional Genomics in plants. She is author on over 73 publications in high impact factor journals including *Nature*. She is a recipient of the Brazilian Science and Technology Fellowship and is a member of the Plant Biosafety Committee in Brazil.

Jorge A. da Silva is a Superiority Professor in the Department of Soil & Crop Sciences at Texas A&M University System. His research include the study of molecular markers applied to plant breeding. He authored over 30 peer-reviewed publications in high impact journals, five book chapters, eight Invention Disclosures and three Plant Variety Releases. He is a member of the Crop Sciences Society of America and the International Society of Sugarcane Technologists.